the logger's Digest

Vol. 1

from horses to helicopters

MCB Communications Ltd.

Credits and Acknowledgements:

The author gratefully acknowledges the contributions of: Leigh Cross, Joy Tataryn, Bob Cain, Kathy Linneman, Mike and Ursula Vaira, Gordy Bateman, C.W. Nash, Pat Rivers, Eva Garthus, Bob Dingsdale, Zbigniew Olak, Jim Challenger, D'Arcy deHart, George McKnight, Brian Chatwin, Elaine and D'Arcy MacKay, John and Pat Casanave, Ted and Sheila Maurice, Harry and Nancy Rai, Chenda Gregory, Ray Seitcher, Gord Barney, Sammy Craig, Gary and Cindi Collinge, Chuck Madill, Doris and Bob Tweed, Hugh and Margaret Gow, Larry and Donna Spencer, George and Annette Shiner, Brian and Betsy Davies, Norm and Susie Kelly, Ken and Winnie Wiegand, Don and Merlyn Kemp, Lois and Andy Bohun, Barry Broadfoot, Byron Johnson, Russel Prediger and Deb Wyndham, Carol Margaret Clutesi, Toni Torgerson, Brian MacDermott, Debra-Lee Massia, George McQueen, Dean Dunbar, Donna Newton, Ray Bloomfield, Suzanne Marion, Peter Robson, Alan Haig-Brown, the Old Buffalo. Special gratitude to Bob Tyrrell.

Also the families: Watts, Soderlund, Devost, Arthurs, Adams, Ellis, Young, Chambers, Kittleson, Haggard, Good, Paul and my family: Sue, Donna, Carla, Marlo, Krista, Brandon, Shawna, Britton and Foster.

Front cover photograph courtesy of B.C. Ministry of Forests. Back cover painting by Bus Griffiths, courtesy of Courtenay Museum.

Front cover typography by The Design House, Nanaimo, B.C.

Editor: Joy Tataryn.

Book design: Leigh Cross.

Printed and bound in Canada by: D.W. Friesen and Sons.

Distributed in Canada by Orca Book Publishers, Victoria, B.C.

ISBN 0-9695896-0-3

Table of Contents

Photo: R. Cain

My Back Pages

" . . . ah, but I was so much older then,
I'm younger than that now."

Bob Dylan, 1969

"Everything is unutterably weary and tiresome. No matter how much we see, we are never satisfied; no matter how much we hear, we are not content. So I saw that there is nothing better for men than that they should be happy in their work, for that is what they are here for, and no one can bring them back to life to enjoy what will be in the future, so let them enjoy it now."

Ecclesiastes

Photo: Courtenay Museum

Introduction: A Dummy in the Crummy

Ever fall in love with your job, or at least have a steamy affair with it? As a kid, I wondered how big people became infatuated with work, and I didn't need to look too far for a classic example. My dad was a logger, and, for many years before New Age consciousness swept in, that particular job title did not require apology.

Today, the logger label paints an image of Hell's Angels with chainsaws loose in our gorgeous forests, getting high on tree killing, while bureaucrats in the background cheer them on. That trite generalization is straight out of the sixties, stuck somewhere between hula hoops, Fabian, flame jobs, and ducktail haircuts. Obviously, loggers of the nineties need more than a New York media consultant to tell their story.

Back in the early sixties, I was baptized into West Coast logging the traditional way–from the snarly end of a one-inch choker. A summer in caulk boots translated into cash for a car, a few

Photo: Z. Olak

boxes of beer, and, hopefully, a date with the cutie next door. So I became a logger–sort of.

The day I turned sixteen, my dad presented me with a hard hat, caulks, lunch bucket, new gloves, and a warning: "The crummy leaves at ten to six–don't miss it!" He said it real slow, more than a suggestion, and all that night I lay sweating in the sheets, praying I wouldn't sleep through my alarm. The prayer worked, because I didn't sleep at all.

Recently, at a twenty-five-year high school reunion, I heard my name being discussed in the bar line-up. My life's progress was rudely condensed by a local redneck lawyer: "Y'know, in all these years that guy hasn't changed one damn bit. He's still just . . . a dummy in the crummy!"

Hmmm, I thought, now, there's a catchy little title, maybe the one I've been looking for. You see, over the last twenty years, I've been gathering these stories, keeping them alive in bars, or retelling them at parties, and it finally hit me–this is my culture, my heritage, my goddamn roots! Or, as Kurt Vonnegut said, "You are what you pretend to be."

And who will care about the old legends after the last crummy is rusting in the marshaling yard? So I decided to jot them down as best I could, realizing that a collection of logger stories by a logger himself may not be everyone's dish of morning medicine.

Call me sentimental, but I lived with those people, and they were salt of the earth. And it bothers me to see them tarred with the same ugly brush John or Jane Q. Public currently brandishes toward the entire forest industry. The facts are: we cut down a whole lot of trees, built a great place to live, shared the booty–roads, schools, hospitals. But we made mistakes. Are the front-line soldiers more guilty than the generals?

Back in 1955, in front of the Sloan Royal Commission on Forestry, H.R.MacMillan lamented the day when British Columbia would be dictated by a pulp economy, referring to the inevitable future transgressors as "fixers with a penthouse view, who never knew rain in their lunch buckets." That noise in the background is

Photo: Bob Dingsdale

old H.R. spinning in his grave.

While I like what the Ecclesiastics wrote, I was never an overly religious sort, but I do thank the Lord for allowing me to be a dummy in the crummy, long as it lasted. All those years were more than just paycheques–or, like the song says, "Those were the days, my friend."

But I still fight an urge to debate anyone who mocks those of us who answered the call of the woods. I guess everyone who belongs to an oppressed minority must deal with similar frustrations. The gang on the crummy will know what I mean. As my grandfather warned us often, "Ya fly with crows, ya get shot at."

In certain incidents that follow, various names and places have been disguised to let me keep what remaining friends I have. You might say these are "made-up stories about real people." But make no mistake–I do not include myself in the pantheon of writers who have mined the rich mother lode of logging history, from M. Allerdale Grainger to Roderick Haig-Brown to Stewart Holbrook to Peter Trower.

But any yarn that wins a free drink can't be all bad–and, now that my power-drinking days are history, I thought I'd set these stories free. At one point, I wanted to present a wider assortment of writers in this collection, but things didn't work out that way. Maybe next time.

An old logger, who only told the truth when he couldn't think of a lie that would fit, said to me recently, "Son, sometimes ya gotta crowbar the facts–otherwise, they don't make sense, know what I mean?"

I went easy on the crowbar. And if these stories got a little twisted in their various incarnations down through the years, rest assured they did happen, more than once, in many places. There are witnesses. ❑

Barry Coulson, 1992
Port Alberni, B.C.

Photo: Author's collection

Photo: B.C. Archives and Records Service

A Matter of Time

Somewhere back in the misty sixties, on the wildest, westest coast of British Columbia, I sweated away an entire summer on the payroll of a gyppo logging company. Any further west and the caulks would've rusted from our boot bottoms. On a clear day we could almost see Japan. In fact, when the tide was high, our tail-blocks were actually bouncing in the goddamn salt water!

Yep, it was a scene Mr. Pablo Picasso would have dug immensely: highball, V-12, tank-mounted, double-clutching, cream-the-bastard, hydraulic, Screamin' Jimmy, load 'n go, rubber-tired, rock'n'roll logging–and, in the background, the whitecaps of the mighty Pacific hissing softly, like a blizzard of doves on blue velvet. Out there on the edge, you might say–in more ways than one.

It was the closest I ever came to actually running away and joining a circus–though it really wasn't much of a run, seeing how my old man owned "the circus." I was a teenaged kid, still green as my grandmother's thumb–and wasting the holidays staggering around a sidehill, inhaling no-see-ems and pulling jaggers from my horny little hands, was not high on my summer game plan. But someone had changed the game.

On the Saturday morning of the July 1st long weekend, I was duly informed by The Landlord, "Son, when this holiday is over, you can start being a logger."

He made it sound like an adventure, like going camping or fishing or out to the go-kart track. Yeah, sure. Thanks, Dad. Sunday evening before my logging debut, with the rest of our family glued to Ed Sullivan and his goofy little mouse pal, I snuck down the basement stairs for a final dress rehearsal. Nine hours till showtime, and counting. Slipping that huge, yellow McCulloch hard hat over my K-Tel Beatle cut; forcing,

Photo: Z. Olak

cursing, cramming my feet down inside those high-top Dayton leathers; a grey Stanfield nicely complementing my new Red Strap pants hooked onto my grandfather's borrowed snap suspenders–man, I figured I was primed and loaded for that mysterious world of hairy-assed B.C. loggers, where men are men and women ride sidesaddle. But, as Grandpa used to remind me regularly, I was just a wee bit wrong–about a whole bunch of things.

Shortly after five the next morning, I received a warning that a whole new set of rules was suddenly in place. Climbing up into the crew bus, I smiled at the driver, whom I'd known forever, and said, "Good mornin', Bob!"

"Oh, I doubt that very goddamn much," he muttered, lurching the bus away in first gear and sending me balls-over-bootlaces down the aisle, toward that lonely place traditionally reserved for the boss's son–the very back seat. Welcome to logging, Slick. Staying long?

Photo: Z. Olak

See, back in those days, the social hierarchy of a one-side gyppo logging show had little to do with seniority and even less to do with registered birthright. The pecking order of a woods-crew put the equipment operators at the very top; the hook-tender close behind; then the rigging rats; next, the native Indians; and, finally, anyone directly related to The Boss.

Obviously, my best chance for survival lay in hooking up with the Indian guys. So I did, almost as if I knew what I was doing. Now referred to as "The Summer Of Love," that month of July 1967 was, as an old donkey-puncher amigo used to say, the longest three years of my life.

At the beginning of his logging career, anyone who could not or would not pack his own weight was quickly canned or chunked out, "run off" and sent "down the road"–or, in the plainest available English even a part-time moron could get, "Your sad ass is fired, Jack!"–with the crew serenading, off-key, in the background, "Yer leavin' . . . on a

Photo: Ted Maurice

float plane . . . "

A body needed to work thirty consecutive calendar days before being officially accepted into the almighty IWA, and, once you were in The Union, you immediately acquired certain rights, which your employer had to honour. But, up till then, you were just another name in the rain—"An' lots more where you came from pal, so if yer gonna fall, fall towards the riggin'—'cause we come to log, not fuck the dog!"

Nobody aimed a word in my direction that first morning until we were all standing around the foreman's pickup, sipping coffee, waiting while the giant engines of the Tyee yarder and Skagit log loader warmed up slowly: electric dinosaurs throbbing in the background. Finally, one of the fallers, Shorty McLean, a guy with a perpetual "hate-on" for the world, said to the bull-bucker, "Whaddya think, Bill—should we piss on this punk, make 'im smell like a man?"

I didn't need a program to know I was their intended target. Then the very large, native rigging-slinger said to me, "So what's your job out here gonna be, partner?"

The beauty of working with native guys was the way they'd automatically treat you like a regular human being, at least till you proved otherwise.

"I think I'm supposed to set the chokers," I told him, trying to sound like I even knew what that meant. He stared blankly at me while the rest of the crew looked away, trying not to laugh, shuffling feet and biting lips in the damp, dawn chill.

"Well, that's a pretty good place to start," he said. "So tell me, where you got your choker hidden?" Dead silence, except for the odd snicker. How could my dad forget to tell me I needed to bring a choker, whatever that was, for chrissake? This was not a great launch for my logging career. I began perspiring in all the wrong places. Places it wouldn't be cool to scratch.

"Aaww, don't worry, kid, you can borrow mine today. Just don't break it, 'cause it's my favourite one." He took off his battered, orange hard hat and conked it down on top of mine, punctuating what he'd just said, though not too hard, and then he winked right at me, while the rest of the crew tried not to wet themselves laughing.

Photo: Z. Olak

I had just survived my first head-on encounter with the great "Ray Seetcher, the chokerman's teacher," as he liked to introduce himself. And, as things turned out, the lessons he was dealing were definitely not available from any books or night school courses. This guy was straight goods.

"This is my cousin, Charlie Jack," Ray said, after the others had all dispersed in various directions, some clambering up into machines. A wiry, little guy with a gap-toothed grin held out his hand.

"How now, partner!" he chuckled.

Then Ray pointed up the mountainside, where a thousand logs lay scattered across each other like a giant game of Pick-Up Sticks. "That's where we work, kid," he said. "Follow me!" And then he was off.

"They've got me confused for a fucking mountain goat," was my first thought, halfway up the hillside, where I'd paused to deposit my breakfast. Through my upchucking, I heard Charlie yell, "Hey, Ray, we must be logging up a bit too high. This young fella's airsick already!"

Eventually, after fifteen minutes of climbing over stacks of jumbled logs and tumbling face-first a dozen times down into hidden pockets of branch-covered holes in the slash, I finally caught up with the guys at the back end. I felt like I'd been in a cat-scrap. From our new vantage point, we could see for miles in every direction: rolling valleys carpeted with dazzling green forest, small lakes glistening in the sun's morning rays–calendar shots whichever way you looked.

Then the serenity of the postcard scene below was rudely fractured by the jangle of the cold, steel butt-rigging. It danced and clanged back to where we, waited perched like a trio of beavers on a giant, downed red cedar. "Bing bing . . . KLANGGG . . . Bing bing bing . . . KERRANGGG."

With his right hand, Ray squeezed the two trigger buttons on "the bug," the electronic radio signal transmitter slung low around his waist. With a screaming "BEEEEEEEEEP" echoing between the mountains, all other sound and motion came to a slow, grinding halt.

"See that?" Ray asked, pointing to the three long cables dangling in the air, a dull-orange blob of steel attached to each one. "The new one on the front is mine, the middle is cousin Chuck's,

Photo: Ted Maurice

and that cute little curly one from the museum is yours. Them's the rules, partner–new guy gets the back bead. Pull 'er hard–she comes easy!"

By the time Ray had shown me a couple of times how to slide the bell up the choker, throw the knob-end across the log, reach underneath, and lock the knob into the hole in the bell-collar to cinch it up tight like a necktie, the novelty of being a chokerman had worn off. Soon, I was wondering out loud when it was time for coffee break.

"We don't give chokermen coffee breaks," Ray laughed. "Too hard to retrain 'em!"

For some reason, I didn't mind their gentle abuse. Twisted or not, it still felt like an olive branch to me. Later, whenever I'd be yanking my choker through the woods and it would snag on a limb or tangle up in the brush, big Ray would yell, "Better pull harder, kid–you're hung up on a leaf!"

If I somehow "mollied" the haul-back block pin in backwards, or was the last guy getting into the clear before he could blow "go ahead" on the turn, or sometimes when he just felt like making conversation, Ray would stare at me and ask, "This your first day in the woods, sonny?" But always with the hint of a smile. Whenever a turn was snorting toward the landing and my snare came undone, or the log slipped out of my bead, causing a dog-cock, Charlie would offer the gentle reminder, "Relax, partner–we couldn't fire you even if we wanted to!"

They used to ask me at least twice a week, "So tell us what you did to your old man that made him give you this wonderful job. It must've been something real terrible!"

At day's end, we'd all race down the hill and stumble into the landing, and old Oil Spill Bill, the mechanic, would shake his head and say, "Man, you three are quite a pair!" I don't know about Ray and Charlie, but I kind of liked it.

Now, all summer long, Charlie had been showing me how he could tell the time by using only his arms and the sun, kind of like a human

Photo: MB Franklin River

sun dial. Whenever the morning seemed to be dragging–pretty well every bloody day–I'd ask Charlie what time he thought it was; if the sun was even partially shining, he'd go into his well-rehearsed routine. First, he'd climb to a spot up above me on the hill. Then he'd gaze directly into the sun for ten seconds or so, whip off his dented blue hard hat, and hold it out at arm's length to block the glare. Next, he'd lift his other arm straight away from his shoulder, causing a shadow on the ground beside him. After studying this awhile, he'd raise one leg up slowly behind him, and announce in the world's worst John Wayne imitation, "It's exactly twenty-seven minutes after ten o'clock a.m. in the morning, Pacific Charlie Time, approximately."

So then we'd blow a long and a short signal on the radio whistle–"BEEEEEEEEEEEP, BEEP"–which meant all the machines in the landing had to be shut off immediately to allow the rigging crew to communicate down the hill.

"What the hell d'ya want now?" the chaser would yell up the hillside, once everything got quiet.

"What time is it, white trash?" I'd holler back. We could see him asking Joe, the grapple operator.

"It's almost ten-thirty, ya jerkass!" the chaser would scream up at us, as the machines roared back to life.

"They're three minutes out down there. That guy must have a Timex," Charlie would say in his laid-back drawl.

Ray would just chuckle. "Relax, partner–we

Photo: Ted Maurice

can always split the difference."

Needless to say, I was heavily impressed the first time they performed this act, as I was every time they pulled it on me—sometimes twice a day, when they were feeling frisky. I couldn't believe how this shaman-logger and his hocus-pocus came within a few minutes of a real clock, every damn time! Soon, I was raving about my pal Charlie's special talent to anyone who would listen.

When I'd ask Charlie to share the secret, his reply was always the same. "Partner," he'd say, "I'll explain it so even a white-boy son-of-the-boss can understand. It's kind of like—magic!"

This little charade went on all summer, till the Thursday of my last week—or "Weekend Eve," as my brown-skinned buds called Thursday afternoons. Finally, I broke down. "Name your price," I pleaded.

Ray and Charlie started jumping around, waving their arms and war-whooping in a zany vic-

Original line drawing by Don Kemp

tory dance, howling with a P.T. Barnum, sucker-born-every-minute kind of glee. The bait was on the hook; the fish was in the mood; it was nolo contendre, Your Honour. And those two merry pranksters were digging their little joke to the max.

"Just a matter of time, partner!" Charlie chortled.

"Aaww, c'mon, you guys," I said, thinking I was close to learning the all-time parlour trick that would help me through college, maybe even to an early retirement.

"Well, son," Charlie said, putting his arm around my shoulder, "I guess the real secret is—the heat." By now, I was in the net; hell, I was almost in the boat.

"The heat?" I asked, trying not to sound too much like a city kid, let alone a size twelve idiot.

"Yeah, the heat," Charlie said, "'cause if you don't get enough heat on the solder, then the damn thing don't stick!" By now, Ray was having a shit-fit, laughing hard enough for three men. In slow motion, Charlie took off his hard hat and held it at arm's length, with the inside facing us. "Took me about three tries before I finally got that sonofagun to stay in there!"

For the longest time, I just stood there, staring at the ancient pocket-watch stuck to the inside of Charlie's tin lid. The damn thing even had a sweep hand. I felt him squeeze my shoulder twice, ever so gently. "Gotcha, kid!"

Ray said later he was going to send in an accident report to the Workers' Compensation Board that said, "Broke two ribs and peed myself while laughing at boss's kid."

The next evening, after work, Ray and Charlie invited me to get off the crummy with them and have a beer down by the river to celebrate my last day on the job. "We've got a real nice reservation for you," was the way Ray put it.

What could I do, seeing as how there was no counter-offer forthcoming from the whiter shade of crew?

So there we were, the three of us, sipping brown barley on the Indian side of the Somass River, when Charlie finally spoke. "Hey, partner, you remember that little trick I been showing you? Well I'd sure appreciate if you kept it, you know, kind of—under your hat!"

Yesterday I went to Charlie's funeral up on the reserve. Later, at the wake, a beautiful, young girl, about sixteen, with Ray's sad, deep eyes and straight, shiny, black hair swinging down below her waist, came across the longhouse floor to offer me a tray of sweets.

"Did you know my Uncle Charlie?" she asked. I smiled at her and reached for a square of Nanaimo bar.

"Yes, I guess I did, a long time ago," I said. Her bottom lip was trembling, and I noticed that she had tucked a well-used hankie in the sleeve of her dress.

"Did you know him quite well?" she asked softly.

"Young lady," I told her, "back when I was about the same age you are right now, your Uncle Charlie and your father taught me how to tell time—the Indian way."

Driving home from across the river last night, I was wishing I could've told her some of the other stuff those guys taught me—way back there in The Summer Of Love, when the three of us had jobs in "the circus."

But, on the other hand, maybe sometimes a guy is better off keeping certain things under his hat. ❑

Dedicated to the memory of Tony Fred, the sweetest left-handed catcher I ever played against, who died in a ridiculous logging accident. He was 41 years old. Thanks for the time, brother.

Photo: Z. Olak

I Have Seen the Eyes of Death

I grew up in a logging camp–if you could call it growing up–and, except for a life sentence in this goddamn wheelchair, I really wouldn't change too much. But that's the joker in the logging cards, I guess. You gotta pay the toll to rock and roll.

Far back as I remember, I've had this weird fascination with power saws. "Motor-cutters," my old man called them. As a kid, whenever I went missing after dinner–always a major concern around a float camp–Mother could usually find me over in the saw shack. The fallers adopted me, and I loved hanging out, watching as they tinkered with their machines, filed chains, and swapped the bullshit. I couldn't wait to get older so I could be a faller, too.

That ramshackle, old shed was the fallers' private domain–their clubhouse–and entry was gained through special invitation only. The rest of camp well understood that shenanigans over in the "Saw Palace" did not belong in the regular world. Out of sight, out of mind, so to speak.

Many years before, some bunkhouse bard had spray-bombed a sign on the door that proclaimed:

What you see here,
What you hear here,
Stays here,
When you leave here!

Back in those creamy times of "big timber, big money," top fallers were in constant demand. They came and went with little warning, like gunslingers hiring out in the Wild West range wars. A hotshot faller was given a wide swath and all the slack he needed. Social crimes and body odour were usually forgiven–if a guy had the magic touch to lay the lumber down.

I was barely thirteen when I finally met Big John Amos in 1968, but I'd been hearing about him for years. If not the best-known faller on the whole West Coast, he was certainly tied for first.

A call from the logger's agency in Vancouver a few weeks earlier had asked, were we hiring any tree-killers? When the boss heard that Big John A. was looking for a place to dry out, he got on the two-way radio-phone in the cookhouse and, with everyone on the coast listening in, told the Mancatcher, "Sure, send that big lunk out here–we'll find a spot for him. Tell him he might have to bunk with the bull-cook for a few nights–that'll sober the bastard up!" I took it they were old friends, Big John and the boss.

For about ten days after that, whenever I heard the sound of McCully's ancient Beaver sputtering up the inlet–usually hugging the deck if old Bill had the stick–I'd be down on the dock before his floats were wet, waiting to grab his wing and hopefully wangle an introduction to this

Photo: Alberni Valley Museum

"Babe Ruth" of B.C. fallers. Finally, he showed.

Now, I'd seen big men before, but I was mesmerized watching that six-six, three-hundred-pound legend gandy-dance up the stiff-leg while packing a huge, new McCulloch chain saw and a large rucksack over his shoulder. Pure ballet. I knew I was witnessing one of the great sights in West Coast logging history—Big John Amos adding another camp to his ever-burgeoning résumé.

Logging camps acquire reputations in many different ways. Great food and clean accommodations help attract a better clientele, while another major factor is the liquor situation. In other words, is the bunkhouse wet, dry, or just damp on occasion?

Because he was still winding down from a "grade-A large twister," as he put it, Big John felt safe in our place. He knew the only booze allowed was what a guy brought with him—inside his belly. By the bulge John was sporting, he was obviously no stranger to elbow-bending.

"Kid," he told me that first week, "I've hunted for the bottom of lotsa bottles in my day, an' you know what? Found every goddamn one of 'em, right where I was looking!"

During his first week in camp, he broke our record for wood-scale that had stood for nearly seven years. As he said one night out in The Pal-

Photo: Alberni Valley Museum

ace, "Funny thing about them trees, boys, as the beavers'll tell ya. Don't matter how damn big at the butt–ya grind away long enough, they all come down!"

But as the summer dragged along, the novelty of being on early shift lost its appeal. We all knew it was just a matter of time before the heat shut us down, and Big John would be long gone.

For some reason, the big guy had taken a bit of a shine to me, pulling me under his wing to share secrets acquired in thirty years of making sawdust, as he put it. "Ya remind me of myself when I was still a punk," were his exact words.

In less than two weeks, he taught me how to spread the tip of a pinched bar with a screwdriver, how to change a starting cord without the spring uncoiling, how to burn off a fouled spark plug, and, best of all, how to file a dull chain.

"Kid," he told me, "you learn how to sharpen a chain so the chips'll fly, I'll guarantee that as long as trees are made outta wood, you'll never want for work."

As I stood beside that giant of a man hulked over his saw, squeezing a rat-tail file in his massive paws and caressing each link like a violinist would a Stradivarius, I knew I was watching a true artist. When Big John was filing, he entered some kind of total Zen trance, completely focussed on making each stroke an exact replica of the one before it, like some ancient religious ritual.

"Push with the back hand, pull with the front," he told me, "an' try to keep equal pressure on both; otherwise, you end up with an uneven gullet. Then the tooth can't take a full bite. Always take the rakers down with a flat file, too; otherwise, they keep the teeth too high in the kerf. I tell ya, kid, goin' up the hill with a dull chain is like goin' downtown with a limp dick. A complete waste of time, an' nobody makes any cash, know what I mean?" He wasn't exactly Winston Churchill, but he got his point across.

Photo: MB Franklin River

Once the obligatory maintenance was wrapped up for the evening and the guys were lounging around the shack, waiting for "mug-up" call, that's when the big fella would trot out his "Bedtime Stories," as he called them–and man, did he have some beauties!

One night, he told us about this little 1-85 float plane he'd won in a poker game from a camp Push, up in Tranquil Creek, which he then traded for a live-aboard sailboat over in Bute Inlet, which he eventually sold to get married down in Vancouver, or "Da Beeg Smoke," as he called it. I was a kind of surprised–he didn't seem the marrying type.

Now, on the few occasions when Big John didn't show up for the nightly bull session, I heard some of the guys make an odd reference to the Port Neville story. They wondered who among them had enough balls to ask the gory details. I was most intrigued. The consensus was that if Big John hadn't brought it up, he still considered the issue somewhat sensitive. Like maybe the sheets were still damp.

So there we were one sweltering Friday evening, loitering outside The Palace, puffing on rollies, when I decided to take my best shot. I figured it was now or never. Any day now, the Forest Service could slam a closure down on us, and Big John would be history. With my little heart pounding like a piston on a three-mile adverse, I waited for a lull in the conversation; then, I pounced.

"John," I said, "the other night a few of the guys were talkin' about Port Neville and wondering if you'd ever been up that far." Everybody got real quiet. Off in the distance, a faint, steady buzz of the dozer-boat was interrupted by incoming waves gently slapping the dump pilings. Sweet fumes of White Sail tobacco bit at my nose as Big John tamped a new load into his famous, old deer horn pipe.

Willie Manson, the bull-bucker's idiot nephew glared at me.

I thought: fuck you and the horse you rode in on. Most of the crew thought Willie'd be dead by Christmas anyway–with a little luck.

"Well, kid," Big John finally said, "just what do ya need to know about that godforsaken shit-hole? As a matter of fact, I wouldn't go back to that rats' nest for a thousand bucks a day, except to maybe piss on the goddamn wharf!"

I thought maybe I'd blundered into forbidden territory, but, as Big John himself once told me, "When yer negotiatin' with a hooker, son, they don't charge for the question. The question is the free part!"

I decided to press on. "Did you really stay in there for eleven months without a break?" I asked, my throat suddenly very dry.

Big John took a pull on his pipe and exhaled long and slow. He turned away to gaze up the inlet as the last blink of sunlight ricocheted off the dozer-boat, still dancing around the bullpen like a lovesick water beetle.

Photo: Bob Dingsdale

"Well, kid," he said, his voice a notch lower than normal, "it's long and ugly, but I'll give ya the Reader's Digest condensed version. First of all, ya gotta understand I was drinkin' at the time. Now, when I say drinkin', I mean like only one or two–weeks, that is. It's hard to keep track when ya ain't wearin' a calendar, know what I mean?"

The big guy took a pull on his pipe, holding it in longer than usual. The rest of us were afraid to move, not wanting to break the spell. John glanced over at me, then continued.

"So I'm holed up for the winter in the Ritz Hotel. I meet a girl, and we fall in love–except for her, I guess. By that time, I was runnin' on straight Seagram's courage, and BAM, just like that, she talks me into a ring job. Am I going too fast for you boys?

"Eventually, after a couple a years off and on, mostly off, she decides the party's over, an' for a cool twenty-five G's, I can be single again! Well, that sounded like a damn fine deal to me, so I packed up my trusty old tree-beaver and headed my ass off for beautiful, downtown Port Neville. Eleven months and twenty-five big ones later, I knocked on her door, we hit the lawyer's office, and to this day I consider that money well spent. Kinda like an investment in my future, so to speak."

The waves slapped louder as the boom-boat headed for the dock. I suppose I should've let that be the end of it, but I sensed there was more. It was almost like he'd left out the punch line, maybe on purpose.

The rest of the guys seemed embarrassed by the personal turn in the conversation. "Sarge" Chatwin, the camp's resident diplomat, not to mention Big John's falling partner, shot me a look

Photo: B.C. Archives and Records Service

that said, "Zip it right fucking now, kid!" Of course, I didn't.

"Eleven months in the same camp, without leaving even once–how could you stand that, John?" I was a nervy little bugger in those days, back before I learned my camp manners.

Big John took a deep pull on his pipe, then turned and shot a long, straight arrow of smoke just past my ear. My eyes were burning. "Sonny boy," he said, icicles dripping from every word, "I guess ya just had to know the woman."

Years later, I heard she'd been dipping the lawyer all along, but who's to know for sure?

Probably the lawyer, for one.

Sure enough, the following morning a call came from the Forest Service to say "the lid" was on, and the whole West Coast was down till it rained. Just after lunch, I wandered over to the faller's bunkhouse to say goodbye to Big John.

"Figgered you'd be around, sooner or later," he said, as he continued stuffing rain gear, dirty work pants, some old Stanfields, and an assortment of books and magazines into a well-travelled U.S. Army duffel bag. The room was musky with the fragrance of old socks, Winter's Green liniment, pipe tobacco, and the history of maybe a thousand loggers spattered against the yellowed, peeling walls. I was mildly surprised to see a Gideon's Bible lying face-down, open on the bed. The Holy Book had done some hard travelling.

"I got somethin' for ya, partner," he smiled, picking up the Bible and taking a tattered envelope from inside the front cover. "When ya knock down yer first pecker-pole, as I expect ya will one day, I want ya to stick this on your hard hat. I hope ya learn what it means, and I hope ya never forget."

I opened the flap of the envelope and took out a brand-new, oval decal with bright-red letters on a green background that said: "I HAVE SEEN THE EYES OF DEATH."

So to make a long story longer, I did become a faller, and for sixteen years I wore Big John's decal in logging camps from Sooke to the Queen Charlottes. Then one sad morning in May 1989, I finally got to know what the damned thing meant.

As the sun peeked through this magnificent stand of old-growth Douglas fir, all green and gold and glistening in the early mist, I remember shutting off my saw to gas up, and then sitting down on a flat, mossy rock for a cigarette. Or, as my falling partner used to call it, "a puff'n'think."

From that marvellous morning all those years ago, when Big John Amos first took me up the hill to babysit him for the day, I'm still not able to breathe fresh sawdust without remembering how he rubbed a handful of it in my face. And how he laughed when he told me, "Kid, that's the smell of new money, and it's the second-finest thing you'll ever stick your nose in–take my word for it. Har, har, har!" I was too young to know what he meant at the time.

Down in a draw at the front end of our quarter, I could barely hear the whine of my partner's new Stihl 056 as he beavered along. I also remember thinking what a wonderful, glorious way to make a living, compared to my kid brother, an accountant. But, as Big John often used to remind us, "Life ain't a contest, boys. Just because yer first across the finish line don't necessarily make ya the winner."

So I spit in my hand, butted my cigarette in it, and cranked up my saw. I had brushed out my runaway trail pretty good, but I remember thinking this old fir I was working on had a fairly bad lean, and if my saw pinched, forcing me to use a jacking wedge, then I just might have a problem. Well, guess what?

The WCB investigation report said that when the tree barber-chaired and brushed the adjacent snag on its way down, a limb came loose in the top. When I ran, I guessed wrong. That's why they're called "widow-makers."

I regained consciousness in the helicopter. After we landed at the hospital, I can still remember how gently the paramedics moved me onto the gurney, ducking their heads under the rotor blades in the grey afternoon drizzle. I also remember the young pilot jogging along beside us carrying my newly dented hard hat, with Big John's decal wrinkled on the front.

About six months later, I received a postcard at the G.F. Strong Rehabilitation Centre, from an address down in Arizona. On the front was an old, white-haired guy sitting in a rocking chair, and he appeared to be smoking a pipe made out of deer horn. On the back was stamped the name "Golden Daze Rest Home," and a barely legible scrawl read:

Heard about your tough luck.
Did the Eyes blink?
Hang in there,

John

The reply I sent returned unopened, marked "Address Unknown," but he still rambles across my mind from time to time.

Last week, my nineteen-year-old son came to visit. He informed me he's moving out from his mother's place this summer, heading up to the Queen Charlotte Islands to work in his uncle's logging camp. We were having a beer—Lucky Lager is all I ever drink—when he dropped the bombshell.

"Dad, what would you think about me breaking in on the saw this year?"

I just sat and stared across my tiny living room at this strapping young man, not believing what I'd just heard. I felt instantly dizzy. I wheeled my chair over to the old hard hat sitting on top of the TV, and I stared at that damned decal for a minute or two, thinking a whole lot of craziness. Finally, I spun around and smiled at him.

"Well, partner," I said, "maybe we should have another beer, and talk about it." ❑

Photo: Z. Olak

Two old hand loggers who hadn't seen each other for several years were bending elbows at a Truck Logger's Convention back in the fifties, when finally one asked, "Say, Pete, didja get that cookbook I sent ya last Christmas?"

"Yep, I sure did," Pete replied. "An' I'm here to tell ya, it ain't worth a pinch of coonshit!"

The other fellow's feelings were hurt. "What's wrong with it?" he asked.

"Well, I can't follow any of the bloody recipes," Pete replied. "See, they all start with: 'Take a clean pan . . . "

1991 LOGGING FATAL

FALLER

AGE: 41 EXPERIENCE: 13 YEARS AREA: SUNSHINE COAST

A faller was cutting standing timber on the edge of an area of blown down timber. As he felled a 2' cedar tree in the same direction as previously felled timber, it appears that a hemlock tree, 1' in diameter, 16' behind the cedar was either leaning onto or was limb tied to the cedar. The hemlock was pulled down or fell when the cedar tree fell. The faller was standing at the cedar stump and had not moved to a position of safety. He was struck by the hemlock tree from behind and killed.

Courtesy of Workers' Compensation Board of British Columbia

Photo: Z. Olak

Photo: B.C. Archives and Records Service

Loggerisms #1

"Take a dose of hike, partner!"
(You have just been fired.)

"Pass me that axe with two heads."
(A city slicker asking for a double-bitted axe.)

"According to Hoyle, in the book of snags!"
(The authority quoted by old-timers when explaining how a job is to be done.)

"Adverse," as in, "That's pretty tough adverse out there."
(Degree of grade against the movement of loaded logs.) ❑

The Wet-Arse Hotel

Photo: Bob Dingsdale

I guess the Eighth Wonder of the World could've been that old outhouse shitter-shack out at Franklin River, way back in the early fifties. At the time, MacMillan Bloedel's Franklin River division had the honour of being the world's largest logging camp, so maybe that crapper was the biggest ever built. You'd have to check the Guinness record book.

It was damned ingenious, the way they had 'er set up, but if someone didn't explain the whole operation to you, the guys would all have a real good laugh at your expense. And you'd end up with a real wet butt.

Y'see, the outhouse was a long shed consisting of ten stalls side by each, and directly underneath was a two-foot-wide flume built on a bit of an incline. Outside the shed, on the side against the bank, was an old, forty-five-gallon oil drum with the top cut out, mounted on a gimballed stand with a counterweight welded to one side.

When the drum filled up with water from the gravity hose out of the creek, it would tip right over, and the water would splash down into the flume, washing away whatever recent "deposits" had been left there.

This whole operation happened every minute or so, and anybody who'd ever used the Damp-Arse Hotel more than once knew the program. Every time that big splash happened outside, it was a good idea to raise your butt up off the seat, or you'd get a blast of ice-cold creek water licking you "right where the sun don't shine."

The old-timers loved to crowd in there when a greenhorn was taking his first dump, because they all knew that shortly he'd be wet and hollering. The night I got nailed, they all enjoyed a

Photo: Alberni Valley Museum

good howl at my swim in the slime, but I guess I got the last laugh. After my own "inauguration," I waited for a couple of weeks to get even, and then shortly after that, the Company decided enough was enough. The brass had a procedure sign painted up to end the shenanigans–but by then, I'd done my damage.

It took about a month before I was accepted as one of the fellas, and I'd worked my way up to second loader, throwing tongs on a track side. So one Friday night, I hotfooted it into the outhouse and smuggled my invention up to the very last stall, right against the water drum.

Now, I'm sitting there with my drawers down, waiting for the suckers to plop themselves in the stalls beside me for their pre-dinner constitutional. As someone invariably used to say, "Out with the old, in with the new."

One by one, I recognized their voices as they bantered away: Charlie Woods, the hooker; Jack McKnight, a truck driver; and Wild Bill Allen, a donkey-puncher. Man, this was going to be perfect–all the old-timers!

Now, all these guys knew about the "stand up once a minute" deal, but they sure weren't ready for my little trick. See, I'd built this contraption out of some old rags soaked in diesel, nailed to a block of red cedar–kind of like a floating Molotov cocktail. So I waited till there were half a dozen guys sitting on their respective thrones, and, when the tidal wave hit the chute, I lit my torch and let 'er go. They all raised up their backsides, but they sure as hell didn't raise 'em up far enough.

"JEEZUS H. CHRIST!"

"SONOVABITCH!"

"WHAT THE FUCK!"

–and then the race was on. I guess those guys chased me for half a mile out behind camp before they finally gave up, and, of course, I had to miss dinner that night, but it was well worth my trouble. I laughed so hard I couldn't have eaten, anyway!

Over the years, that particular prank has earned me a lot of free drinks in exchange for the story. But that wasn't the end of it.

In 1955, at the Company Christmas party, I was appointed Lifetime President of the Franklin River "Burnt Ass Club," complete with a huge trophy to prove it. The trophy is a miniature butane

Photo: Alberni Valley Museum

Building Bearshit Canyon Trestle, Franklin River.

Photo: R. Cain

Things are different nowadays.

torch with a welder's flint to light it, all mounted on a gorgeous cedar burl–complete with some burnt hair glued on top. And the damn thing even floats! ❑

I first heard this story from the late Rocky Grant back in the early 1970s, and later details were verified by Ted Maurice, now retired and living in Nanoose Bay.

A logger was making fun of a shepherd sitting next to him at the bar. Finally, the shepherd could take no more."You doggone loggers think you're so smart and tough, and maybe you are. But when we're herding sheep, we do all our work with our heads higher than our butts. I got no respect for a man who works with his hind end higher than his head!"

A Logging Accident

Leigh Cross

Photo: Z. Olak

We didn't notice Harry, the new chokerman, was missing until we sat down to dinner. "Well, we got to go look for him, that's all," said the Push.

"Shit," said Hank, the loader.

"Shit," thought all the rest of us. We were hungry as hell.

"Let's go," said Big Charles, the hooker.

Dinner would have to get cold. We all got up from the table, the whole goddamned crew. Some of the guys slapped a steak between a couple of slices of bread to bring along with them, and the cook and the bull-cook took the other platters of food off into the kitchen to keep them warm. We grabbed a few flashlights, put on our working clothes, still slimy from the day's work, stuffed our feet into our caulk boots, and piled into the crummy. The Push drove.

Photo: Author's collection

Big Charles snarled at everybody; he felt lousy; it's the hooker's job to keep track of his men. Matter of fact, we're all supposed to look out for each other, but breaking in a green man is such a pain in the ass that we didn't exactly notice when Harry disappeared, any more than someone notices the exact moment when a toothache disappears.

The Push drove fast. The crummy rocked and bounced as we climbed the rutted and pocked logging road through acres of slash. We gritted our teeth and hung on to whatever we could grab. The moon, a lamp over an operating table, lit up all the gashes and bruises on the hillside, stumps, snags, clots of tangled, smashed branches

Photo: Ted Maurice

and small trunks–all the offal of logging.

We reached a level spot where the road ran through one of the few remaining groves of standing timber: primeval old-growth, black as death, where the Tree God still lived and hated. His clammy breath coiled and misted in through the windows of the crummy.

We closed the windows. There's no such thing as a Tree God, anyway, bunch of dumb-ass bull-shit, that's all. Logging's a job, maybe a little more dangerous than most, but only a job.

Photo: Bob Dingsdale

Photo: Author's collection

The crummy bounced and rattled around a bend and back into the moonlight. More slash. Everyone's eyes were shaded by their hard hats and I could only see mouths. The guys with the steaks gnawed and munched away. The blood and juice ran out of the steaks, over their hands and down the front of their shirts. "Christ, what a mess. You guys look like you'd died in a knife fight," said Big Charles.

Hank had his mouth full of steak and bread. He said: "What's the difference? We're going to have to hose the gore out of the crummy anyway, one way or another."

"Shut the fuck up," said Big Charles.

The crummy pulled up and stopped at the landing by a deck of skinned and battered logs and the steel-spar, a fat, ninety-foot piece of steel pipe poking insolently, straight up into the sky: a rusty, yellow prick to fuck the Tree God. It's risky, throwing a fuck at God, but that's logging.

Photo: Author's collection

Photo: R. Cain

We formed a ragged rank about fifty feet wide either side of the butt-rigging where it sat on the landing, Big Charles at one end, the Push at the other. "Keep spread out like that," shouted the Push, "but stay in line. Keep track of each other. Don't get ahead or behind. I don't want to lose anybody else. When the moon goes under a cloud, stop. No use moving if we can't see, OK? And keep your ears open and your traps shut. Maybe Harry's OK. He could only be hurt, you know, hurt, that's all. Maybe we'll hear him; maybe he'll hear us and yell."

We moved slowly up the side of the hill under the stretched rigging. We passed the flashlights back and forth to search the shadows around the stumps for a man smashed into a soggy lump by a log crashing down the hillside.

We peered into each tangle of slash for the shredded rags of something that might have been tossed there. When a diesel yarder stretches eighteen hundred feet of haul-back wire singing tight, it can snatch a man up and spin him maybe a hundred feet through the air, ass-over-teakettle.

We probed each slimy, marshy spot to see if something had been ground down into the mud.

Every so often, we'd yell Harry's name and listen. Not a sound.

Why did it have to be me that walked on Harry's face with my caulks? Not that it made any difference to Harry. ❑

Photo: Author's collection

Photo: Alberni Valley Museum

"It will be a sorry day for British Columbia when the forest industry here consists chiefly of a very few big companies, holding most of the good timber—or pretty near all of it—and good growing sites, to the disadvantage and the early extermination of the most hard-working, virile, versatile and ingenious element of our population: the independent market logger."

From an address by H. R. MacMillan to the third Royal Commission on Forest Policy, chaired by Chief Justice Gordon Sloan in 1955.

Photo: Z. Olak

Bar Vibrations #1

Unseemly chatter in the bar of a chainsaw or, on the other hand . . .

There are worse places to spend a Saturday afternoon than the Rodmay Hotel beer parlour–though not many. Quite aware of the hard-ass reputation this Powell River pub has earned throughout its seventy-five years, I tried to stay low-key. From a corner table beside the old, bubble-style Wurlitzer, I recently spent several hours as a fly on the wall, overhearing some of the more amusing conversations I've enjoyed in many years of bar-storming.

"So I hear the Company's latest line of crap is job-sharing. That's fucking Communism, man, look what happened in Russia! The Bible says everyone's equal, well I gotta call bullshit on that one, partner. We all get twenty-four hours in a day, and equal stops right bloody there!

"I'm up at five-thirty every morning, catch the crummy at six, I get home twelve hours later dragging my ass from a day in the brush, and I make a hundred and fifty bucks a day. Now, how can you tell me that head guy on those TV commercials, or whoever runs the company now, deserves to pull in six hundred grand a goddamn year with all his bonuses and stock options, when it's the boys in the crummy doin' all the dirty work?

"Who're the poor buggers stuck out in the rain and snow and heat and bugs and early shift, and who gets laid off a couple of times a year while the office staff get to sit around in their candy-ass air conditioning playing with computers? Then they parachute in these high-priced efficiency experts straight outta college who tell us we gotta make cutbacks and kick up production, and then they go and build a new mill in Alabama, for chrissakes, with all the goddamn profits we made for 'em up here. It's bullshit, I tell ya!

"And how many visits to the hospital do those office jerks make in a year, or how many funerals of buddies killed on the job do they have to go to? I mean, is sitting behind a desk all day a high-risk situation? Does a computer kick back, or a calculator catch fire, or a photocopier roll over and jack-knife?

"Yeah, I'd like to see the WCB fatality statistics on management versus the labour crew. And then they rap off about team-work and Company pride. Maybe it's time the team made a trade or two, or fired a few managers, 'cause the players are working hard as we bloody well can!

"And where's management every time we have a demonstration to tell Victoria it's real people losing our jobs, not just numbers? It's us loggers on the front lines every time taking all the media flak, like we're the bad guys, when really we're just one small part of the whole damn chain. If my job goes, then so do another two around town in the support industries, but people haven't wised up to that yet. It's the old domino effect, just like Ocean Falls or up in the Queen Charlottes when they made Moresby a park.

"When the logger's hit the road, there ain't much reason for people to hang around, and then we'll see how tourism is just a part-time deal. What about the other six months a year? Yeah, you could say I'm a little pissed off, and who's gonna fight the forest fires every summer? Not the tourists or the tree-spikers, I'll bet!" ❑

Photo: B.C. Archives and Records Service

Photo: Z. Olak

Native Justice, 1969

Bob Fitzgerald was possibly the best damn crummy driver on the whole West Coast. A cat-skinner with thirty-two years of roadbuilding under his belt, "Uncle Bob" had a gruff, surly streak, though I think some of it was just for show. He loved chain-smoking those godawful White Owl cigars, and at six o'clock in the morning, mixed in with various other logger stenches, that aroma bordered on disgusting.

But regardless how wild the boys had partied the night before, if they made it on the bus in the morning, they could grab an hour's power snooze and wake up on the job. That was the beauty of Uncle Bob. In twenty years at the helm, that old buzzard rolled our crew a million miles or more, entirely accident-free.

Oh, there were a few incidents along the way, of course, but nothing too serious. Except for the Friday night that "Snuffy" Arthers sat on a mickey bottle, carving his butt so badly the Company finally nixed booze on the bus. While the wives applauded that decision, the crew was a tad pissed off. Friday's ride home was never the same after that.

Uncle Bob considered his position one rank beneath God. On an Intensity Exam, he would have ranked up there with brain surgeons or rugby players. Many evenings, a crew member would phone Bob's house with a message, only to be informed that His Nibs had just gone to bed. At eight o'-fucking-clock, no less! Most of the younger guys were still climbing into their town duds, headed for a shift in the nearest bar, and Uncle Bob was flat in the rack already.

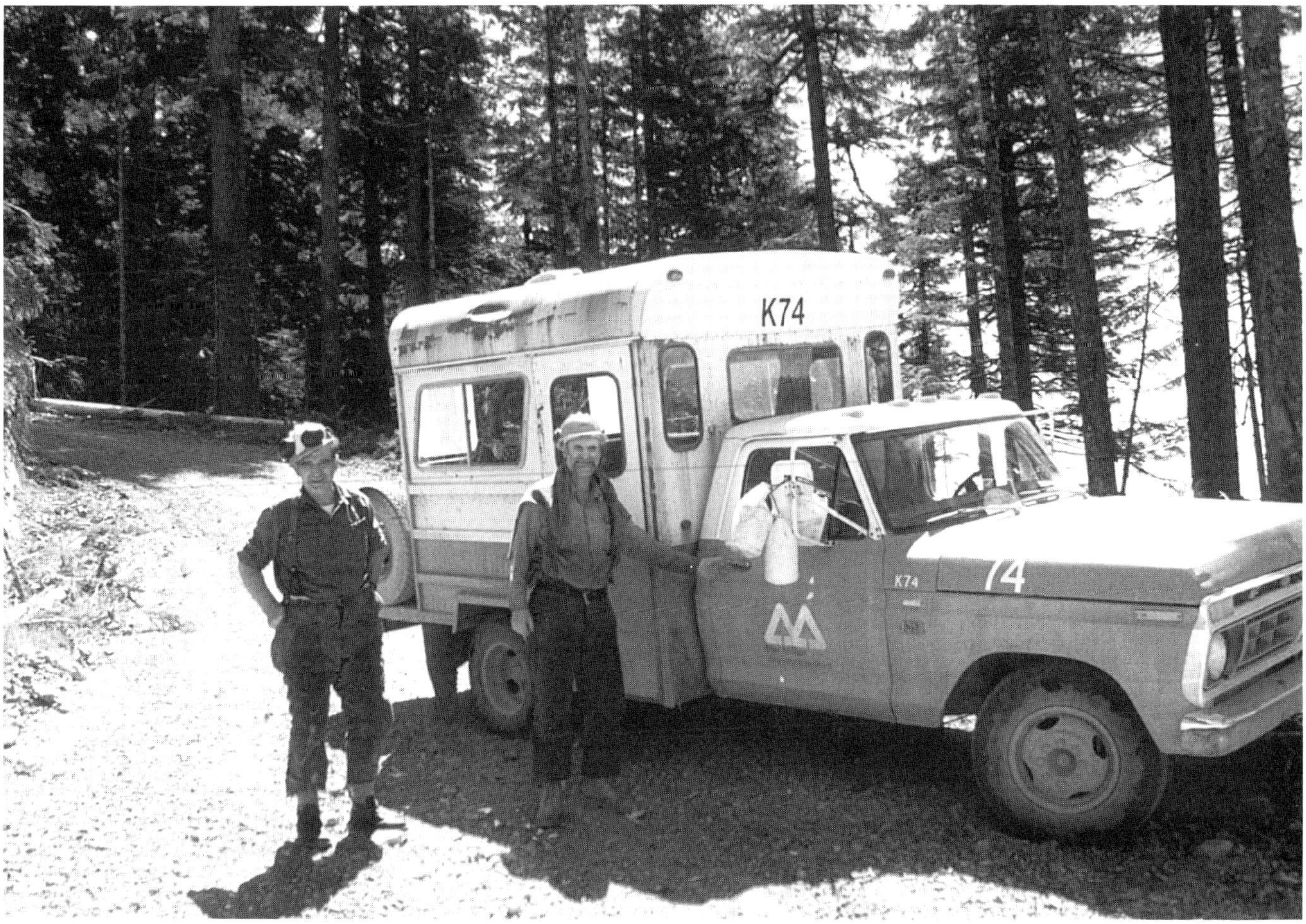

Photo: MB Franklin River

Photo: MB Franklin River

We never worried about him dozing off at the wheel, but if you got on his bad side, then you had a problem. Uncle Bob didn't really need a legitimate excuse to dislike you–any vague reason would suffice. He didn't like me because I was a snot-nosed, teenaged punk, the boss's son, and worst of all, a Montreal Canadiens fan. He didn't like Norm Mack because Norm was an Indian, and also the hook-tender, a very good one.

Norm and I both learned the hard way about Uncle Bob's double set of rules for picking up the crew in the morning. If Bob liked you, he would park at your loading spot and wait there five minutes. Sometimes, if he thought you'd slept in, he might even send a guy up to bang on your door.

But if you were on his shit-list, and weren't standing out on the street where he could see you, then you were S.O.L. The old bugger would barely slow down, shift gears, and then disappear even before you had a chance to run from the porch or wherever you were hiding out of the rain.

Not only did you then blow a day's pay–you also faced the foreman's wrath and the crew's razzing when you finally did show up. Bob had pulled this twice on big Norm, and what later transpired was my first experience with good, old-fashioned, West Coast Indian justice.

Back then, our crummy was an ancient, beat-up GMC two-ton with a cold, steel box on the back. Supposedly, it seated ten, though never comfortably, and it had a buzzer-line to signal the driver, same as a regular city bus.

Because of Bob's stinky cigar, no one would ride up front with him in the cab except the powderman, who was a peaker short of a full load at the best of times.

On one particularly ugly Monday morning in November, Bob pulled the bus up in front of the grapple operator's house and found the place in total darkness. No sign of life whatsoever. Jim had slept in again, so Bob waited with the motor idling. After a few minutes, Norm reached over, opened the side door, and slammed it hard enough for Bob to hear up front. Norm clomped his funky old Romeos across the wood floor toward the buzzer-wire, pulled it three times for the "go ahead" signal, and calmly sat down.

Bob figured his man must have climbed aboard, so he pulled the bus away and headed toward Sproat Lake, where we usually picked up the last guy, Randy Lamb, who also happened to be the yarding engineer.

When we arrived at Randy's driveway, it was the same scene as ten minutes earlier–raining like a bastard, the place totally black, no life anywhere. We all knew how much the two engineers loved their booze, and figured they'd climbed into a Sunday night bottle of rye. Once again, Uncle Bob parked and waited for his buddy. After a minute or two, Norm pulled the same sneaky routine, and Uncle Bob tumbled for it twice in a row.

Finally, out on the highway heading for Canoe Creek, the boys in the back were pissing themselves. We knew the shit would hit the fan when we arrived at work with no one to run the machines, but we all wanted to see the look on Bob's face when he had to tell the boss.

By the time we hit the gravel cutoff and headed up the hill, the rigging crew had agreed that if anyone got fired over this, we'd all quit. Daylight was just breaking as the crummy pulled into the landing, nose to nose with the first truck already backed in under the grapple, ready to load.

No one made a move. Up front, Bob gunned the motor to hustle us up. We all just sat there. Suddenly, the door opened. Bob poked his head in and counted four bodies instead of the usual six. "What the fuck's goin' on?" he snarled. Dead silence. Norm just stared at him.

"Quite a bit of rain, by the looks of things, partner," he said, finally. It wasn't easy keeping a straight face.

"Where's Jim and Randy?" Bob asked, trying his damnedest not to sound nervous.

Norm took his time lighting a hand-rolled cigarette, holding the match till it almost burned his fingers. "Looks like you forgot to pick 'em up, old-timer."

Bob knew he'd been had, big-time. He threw his cigar on the ground and screamed at Norm, "I'll have you fired for this, you goddamn Siwash asshole!"

Norm just smiled. "That's OK, partner. I was lookin' for a job when I found this one!"

Normally, the trip into town took ninety minutes, but we made it in half that, and I think old Bob hit every pothole he possibly could, just to spite us. When the bus finally made the pavement, we were all hurting pretty bad, mostly from

Photo: MB Franklin River

laughing. Bob drove straight to Company headquarters without stopping to let any of us off at our places. He left the crummy running, and we watched him charge up the office stairs two at a time, mad as a drunk who'd missed last call.

After five minutes or so, Uncle Bob and the boss came out toward the bus, and we all figured the party was over. The side door opened, and there stood the Old Man, trying to cover his smile with a frown, though not quite pulling it off.

"Howdy, boys," he said. "How's she goin' this morning?" Four pairs of feet shuffled nervously.

"Oh, she's goin' . . . barely," the chaser answered meekly.

The boss unwrapped a piece of gum, usually the sign he was going to can somebody. "Almost ten o'clock–looks like we're gonna have to take the rest of the day off, if that's OK with you, Norm." And then he winked.

Norm leaned forward and looked the boss straight in the eye. "Don't you worry, Cliff, I'll speak to that guy up front so it don't happen again." With his deadpan face, Norm would have been a helluva poker player.

The Old Man finally broke into a small grin. "Oh, I doubt that'll be necessary, Norm. I think Bob's already got the message . . . from both of us, if you know what I mean."

About six months later, Norm slept in. It was the morning before Good Friday, which meant he'd lose his paid holiday if he missed the crummy. I was alone in the back of the bus when Bob pulled over at Norm's place and gunned the motor a few times, the usual signal the crummy was waiting. Finally, the side door opened, and there stood Uncle Bob, rain dripping off his greasy Cat hat.

"I guess ya better run in there and tell that guy he's got two minutes max, or we're outta here!" Even though it was still dark, for a second there I thought the old bird was almost smiling. Then again, with the rain coming down, and that damned cigar smoke, and my hangover, maybe I was just seeing things. ❑

Norm Mack and friend, 1992.

Photo: B.C. Archives and Records Service

Loggerisms #2

"Alibi day."
(Payday in camp, when excuses get invented to head for town.)

"It's a barb-wire deal!"
(A tough situation to handle.)

"He's got the old barleycorn sprints!"
(Describing bowel fluctuations after a booze session.)

"It's kind of a bastard size!"
(Anything peculiar or off-sized in equipment.)

"He's a pretty fair bean burner!"
(Describing the camp cook.)

"We're gonna high-grade that next setting!"
(To take the best trees only from a stand.)

"The doughnut holes are too damn big for me!"
(Said of a camp where food was skimpy or pay was cheap.)

"He's checking out the daisies from the root end!"
(Describing someone who has died.)

"That guy is looking for a stretcher ride!"
(Describing a man who takes foolish chances or indulges in careless horseplay.) ❑

Photo: R. Cain

The Shop

Photo: Courtenay Museum

In any logging company, there is no place more important than "The Shop." It is the nerve centre where new ideas are hatched and developed, the hospital where sick machinery is taken to be revived, where mechanics burn the midnight oil trying to revive equipment long overdue for the boneyard. It is a place where magic happens every day.

Because new logging equipment has always been painfully expensive, the forests have long been littered with antique machinery. However, with a shot of T.L.C. from the master mechanics and an occasional prayer from the owner, some machines pay for themselves many times over.

The shop foreman, usually a combination of welder-mechanic-inventor, wears a status similar to

Photo: Courtenay Museum

The "Friday" hammer:
The biggest hammer in the shop.
After a few swings, you felt you'd been swinging it all week.

that of a head surgeon in a real hospital. His responsibility is huge. Not only must he give immediate attention to whatever is "broke down," he is constantly thinking about P.M.–Preventive Maintenance. How to keep all the rides in the circus going up, down and around–and stopping only when they're supposed to.

The modern shop has come a long way from those early days of a canvas tarp stretched over a wooden A-frame, when the bone-chilling cold could easily freeze a wrench to the skin of a hand. Today's shops have the most modern conveniences: heated floors, halogen lighting, power tools, computers to keep track of parts and

Photo: MB Franklin River

Photo: Bob Dingsdale

Photo: MB Franklin River

pieces, weekly cover-all service, and outrageous calendars in the lunchroom.

Mechanics are a most unique breed, with their "if you can break it, we can fix it" attitude. From experience, they know that even a brand-new machine will develop "bugs," and these flying doctors with tool kits for black bags take great pride in knowing their home phone numbers are always high on the Company list.

Listening to mechanics swapping ideas in Shop-ese can be as intimidating as watching a movie in Swahili. "Say, Werner, I wonder if we couldn't yank those old friction blocks out of the Madill we junked, and maybe shave the ends so they'd drop backwards into those casings Walter saved from the Tyee–redrill 'em, of course–and even though we'd lose some grab, we might sneak by till the weekend, and then swap the whole shitteree when we flop on the new Gearmatic–whaddya think?"

"Hell, let's try it–she ain't gonna fix herself!"

Around most logging shops people operate with the idea that if it hasn't been attempted, it probably hasn't been thought of yet. Mechanics love swapping stories of how they "gerry-rigged" a piece of defunct machinery and blew productive life back into it, just to finish the day, or week, or till next time it breaks. To shop mechanics, a busted piece of equipment is a challenge, a game, fun.

As old mechanics used to say about running equipment, "Two things can happen, son. She'll get better, or she'll get worse–but if it ain't broke, then don't fix it, 'cause that's my job!"

Rather than get despondent over the never-ending breakdowns, mechanics accept each catastrophe as a part of the territory, and their wives know all too well about that dreaded phone call: "Yeah, honey, it's me–I guess I'm gonna be a little bit late . . . You're kidding! . . . Midnight already? . . . Well, we're just gonna pop the main shaft back in and I'll be right there . . . The weekend? . . . Uuhh, we'll have to talk about it later . . . Hello? . . . Hello?" ❑

Photo: Z. Olak

Photo: Bob Dingsdale

Three adventurous women from the big city decided to spend their vacation at a dude ranch. The first morning was spent admiring all the strapping young cowboys.

Eventually, the bravest of the three friends approached one of the hired hands. "Gee, that sure is some hat you're wearing," she asked. "but tell me, why does it have such a huge brim on it?"

"Well, ma'am," the cowboy replied, "that's to keep the sun out of my eyes when I'm ridin' the range."

She then enquired, "Now, that gorgeous neckerchief—"

"Well, ma'am, that keeps the dust out of my throat when the trail gets really dry."

The woman didn't stop. "I see. And what are those leather things you have on over your pants?"

The cowboy was very patient. "Well, those are called chaps," he told her. "They keep my clothing from getting ripped up when I'm riding through the brush."

Still she continued. "OK, then, what are those little silver things on the back of your ankles?"

"Spurs," he replied. "We use those to control the horse, let him know exactly what we're talking about."

"Boy, that makes sense," she said. "Now, I understand the hat and neckerchief, the chaps and spurs, but one thing puzzles me. Just why are you wearing those funny, white tennis shoes?"

"Well, ma'am," he chuckled, "that's so we can tell us real cowboys from the logging truck drivers!"

Al West

Al West is eighty-eight years young. He began his logging career in 1918 and was the general woods foreman for MacMillan Bloedel's Sproat Lake division when he retired in 1965. Today, he still keeps busy around the remodeled Camp One bunkhouse that he and his late wife Alma moved onto their country property that same year. Al bakes a mean batch of cookies, tends a fairly large garden, and loves driving down into Washington State to visit his grandchildren. A personal friend of the late H.R. MacMillan, Al has many fond memories of their hunting and fishing trips, and loves recounting stories involving the Old Man. Following are two anecdotes he told me during the winter of 1991.

Photo: Alma West, 1949

This picture of Al stood on H. R. MacMillan's desk for many years

Photo: Alberni Valley Museum

Early one Sunday morning back in the early fifties, I was taking H.R. and a few of his cronies out deer hunting in the Ash River area. Of course, it was still dark when we all climbed aboard the speeder, and on this particular day, I was a bit anxious because my fourteen-year-old son was along for the first time.

He'd spent the summer punking whistles, but he wasn't too happy about it. Seems that the manager, Don McColl, was paying him ten cents below the regular rate for the job. Even though my boy was doing the work just as good as the other punks, he still wasn't sixteen, the legal age.

So we bounced along the tracks for about fifteen minutes, everyone still a bit sleepy, when H.R. leaned over and asked my boy, "Tell me, young fellow, are you going to be a highball logger like your dad when you're older?"

My son didn't miss a beat. "Well, if I am," he said, "I sure as heck don't want to work for this cheap outfit!"

I couldn't believe my ears. Either he didn't know who he was talking to, or he didn't much care. I stepped on his foot immediately, hard, hoping that would shut him up. H.R. saw my concern, but wanted to hear more. "And just what do you figure is wrong with this company?" he asked.

My son proceeded to explain his side of the situation, and how unfair he thought it was. The Old Man listened with a bit of a frown, and then he said, "Well, I want to thank you for speaking your mind, young man, and I hope you'll give this company a second chance."

On his next paycheque, Roy got the regular whistle-punk rate, and two weeks later, the manager in question was transferred up to the Queen Charlotte Islands, with no explanation. I don't think it was a coincidence, either. ❑

I guess because I live out here alone, my neighbours were getting tired of checking up on me all the time. So finally they asked if I'd mind going on this new Lifeline program.

It's a pretty good deal, actually, for old people. They give you this little buzzer to wear around your neck so if you fall down or need help, you just press the thing, and a signal rings on the alarm board in the hospital. Then the hospital can see who's in trouble and call one of the four neighbours that are on your helper-list.

Photo: B.C. Archives and Records Service

Photo: *Author's collection*

Al digs in his garden, 1992.

So last week I was getting a little firewood up in the back of my property when I noticed young Rob Hill come charging up the driveway, shirt half-buttoned, looking like he'd just woke up. Those pulp mill guys work haywire shifts, anyway, especially when they're loading a ship.

I believe that was the first time I'd actually seen big Rob running, so I figured something serious was up. I was halfway through the second cut on this slabby old cedar I'd knocked down when I shut the saw off to hear what all the commotion was about. "What's the problem, Al?" the young fella puffed, all red in the face. I assumed he was asking about my bucking ability.

"Oh, I think my chain's a little dull," I told him. "I probably rocked it, chewing on that butt cut."

"No," he said, "what's wrong with *you*? The hospital just called, said your Lifeline went off!"

Then I realized I must've accidentally pressed on it when I was leaning over the cedar with the saw, trying to get more bite before I hiked around the other side.

Rob was chuckling pretty good. "What am I gonna tell the hospital?" he asked. "The old guy's OK, but his chain needs filing?"

"Sure, that sounds pretty good," I told him, "but maybe don't mention that I rocked the saw, or before you know it they'll have a bull-bucker out here giving me a pep talk."

As I cranked over the old McCulloch, I watched young Rob heading down my driveway, shaking his head and doing up his belt. We got some pretty darn good neighbours out here. ❑

Overheard at a barn dance in Union Bay, as two women were entering, one obviously from the big city. "Now, Clara, you be real careful tonight. When some of these loggers get dressed up, they almost seem like regular men!"

Photo: B.C. Archives and Records Service

Loggerisms #3

"Looks like we gotta make a meat run!"
(A special trip to haul out injured or dead men.)

A "misery harp" or "misery whip."
(A crosscut saw, and those who don't think so have never pulled one.)

"The monthly insult."
(The paycheque.)

"He got no place to stood his foots!"
(When a boom man jumps to a log that won't hold him up.)

"Here comes the brains!"
(The head man, boss, or owner is arriving.)

"That guy is just burning daylight!"
(A logger working with little progress.) ❑

1990 LOGGING FATAL

WORKERS' COMPENSATION BOARD OF BRITISH COLUMBIA

SKIDDER OPERATOR

AGE: 26 5 YEARS EXPERIENCE AREA: SOUTH KOOTENAY

As the skidder started to slide sideways down the slope, the operator attempted to jump out of the up-hill side of the skidder. At the same time the main line snapped and the skidder violently rolled down the hill. The operator was thrown out and crushed by the canopy.

Courtesy of Workers' Compensation Board of British Columbia

Photo: *Larry Spencer*

Photo: *R. Cain*

Bar Vibrations #2

Photo: Alberni Valley Museum

Well, ma'am, that man's the—ah—um—the hooker.

"First time I ever met Harry was in the cabaret of the Discovery Inn, over 20 years ago. It was Girl's Night Out for me and my two sisters, and these five guys rolled in acting like they owned the place.

"Apparently, they'd just chartered a plane down from some camp after a twenty-one-day stint, and were looking to party hearty. My oldest sis thought they looked like the Willie Nelson band, with their long hair and ball-hats and red-strap jeans hacked off six inches up their ankles, but they were loggers, and kinda cute, actually. Pretty soon Harry and his buddy were over at our table, asking did we want to buy them a drink, otherwise they wouldn't go home with us! These guys definitely had nerve, but they were sort of funny, so after a while we got up dancing, and the next thing you know, we were all out having Chinese food. To top it off, our first grandchild was born a month ago today!

"I think being married to a logger is like marrying a policeman or a musician–you share him

Photo: *Alberni Valley Museum*

with the job. The one thing I don't like is when he's late getting home, for whatever reason, because I always think the worst. It's a fact of life that people get hurt in the woods every day, so whenever I hear an ambulance siren or a helicopter coming in, I say a little prayer and hope it's not Harry.

"I often wonder if he'd be happy selling cars or driving a school bus, but he was a logger before we met, and his dad was, too, so I guess it's in their blood. Even though he's always complaining about the heat or the snow, the foreman, the crummy ride, his wages or his old machine, you should hear him freak out whenever the preservationists are on TV talking about stopping all the logging. I usually send the kids downstairs.

"I say to him, 'Now honey, do you use language like that in front of your logger pals?' And then he laughs loud enough for three normal men, just like the night I met him." ❑

Photo: *MB Franklin River*

Photo: B.C. Ministry of Forests

The Decision

I fly helicopters for a living. At least, I used to, up till yesterday, when the shit hit the tail rotors, as the engineers say around the hangar. This morning, the desk jockeys at the FAA made a decision to lift my licence "indefinitely." Not for thirty days, or a year–indefinitely. Man, talk about your biological clocks!

As of nine a.m., July 3rd, 1991, I've been officially set down, grounded like a wayward teenager. "Unfasten your seat belt, Slick, your torque has been took. We are parking you awhile." And all because I made one wrong call–at least, in somebody's opinion. Maybe I have an attitude, but if that situation comes up again, God forbid, I know I'll do the same bloody thing–even if they yank my wings forever. Forever I can relate to. Indefinitely confuses me.

Wouldn't you think a record of nine thousand hours in the air with a pair of sixty-foot blades zipping past your earballs, and not the slightest smudge of an incident might impress the briefcase boys? Apparently not.

So how did I auto-rotate down into the chicken-shit soup, you wonder? Ask me an easier one. The official report will say I went against the book, and that's a no-no. Maybe the book needs a little editing job. Maybe if I'd said I had a 'Nam flashback, like some guys do, they'd have slapped me a little softer. Unfortunately, I didn't think about it till just now, halfway through this jar of Mr. Walker's reddest.

About an hour ago, I called my old instructor, who babysat me the very first time I took the stick of a bubble-nosed Hughes 500, down in Medford, Oregon, circa 1975. Harold is the Babe Ruth of Sikorsky heavy-lifters, and even though he's been officially retired for ten years, his phone never stops ringing. Heli-companies everywhere ask him to train their hundred-hour hotshots, engineers love picking his brain, and all the guys he flew "left seat" with still check in periodically, if only to wail their latest tale of woe.

Harold told me I made the right move, so that took away some of the sting. "Don't let the bastards grind ya down, son," he said. He also suggested I go public with my story.

Photo: Author's collection

"Harold told me I made the right move."

Three years ago, I was locked in as junior man on the roster of heavy-lift captains for the world's largest heli-company. I loved everything about the job, except that all nine guys above me loved their jobs, too, and ten years up the line we'd all be in the same positions. I was frozen tighter than a polar bear's butt.

Then came the phone call that changed everything, the old "offer I couldn't refuse" trick. Seems these guys had been logging for years out

Photo: Larry Spencer

" . . . hooking under a helicopter like working inside a hurricane . . . "

on the West Coast, but were fast running out of accessible timber. Over dinner and a few beers, it came out how the Old Man had pioneered a lot of innovations in conventional logging; now, the son was picking up the torch.

Seeing as how my own dad had been a high-rigger and a gyppo himself, back in the '50s, I felt real comfortable with these two. They had been working on an airborne grapple concept in their shop for about a year, and now they wanted to try it out. They wondered, was I interested in being the first to fly it? Does Howdy Doody have wooden cojones?

We took a few months shaking the bugs out, but Rome wasn't built in a weekend, either. In the old days, before the grapple, the game of helicopter logging was somewhat dangerous. Getting a ten-ton machine to park in the air long enough so a two-hundred-foot cable underneath stops swinging, and holding it steady while some poor schmuck hooks up the turn, ain't no easy trick. A good pal of mine once nailed a kid on the ground and was never the same pilot after that. He's a car salesman now.

The downwash alone from three thousand horsepower makes hooking under a helicopter like working inside a hurricane, or an expensive version of Russian Roulette. With the invention of the grapple, we changed the game completely. Safer, faster, cheaper. As Jim Ryan, our chief engineer, says, "Lots mo' bettah now, Jack!"

But even though we've taken logging to the ultimate with this giant set of hydraulic tweezers dancing out of the sky, we still have one big problem. We still need fallers. From the outset of any logging show, the dirtiest job on the claim has to be the guy with the saw.

As chief pilot, I had the duty of checking out the variables of each new site, and then deciding how to proceed. Landing sites, approaches, windshear corridors, abort zones, falling quarters, and where to build the heli-pad were all decisions in my responsibility.

Photo: Author's collection

" . . . in the middle of nowhere . . . "

Building a landing pad in the middle of nowhere is a highly skilled art that lives depend on, including my own–so naturally, we go with our best. Joe MacBurney has been felling trees for thirty years on the B.C. coast, without a lost-time accident. That fact alone tells more about him than he ever would. The guy definitely knows his onions. Big Joe has been my main man on the saw ever since I joined this outfit.

"Bobby," he once told me, "it's not that I got thirty years' experience. I basically got just the one year's experience ... thirty times over!"

In the three years we've had our team together, I've come to trust Joe's judgment whenever it gets down to the real short and curlies. As the guys in the crummy say, "Ol' MacB.'s still got his no-hitter happening."

Photo: Cindi Collinge

" . . . giant pair of hydraulic tweezers . . . "

One night, about a month ago, he came over to my room for a chin-wag after dinner, which surprised me a bit. Usually, Joe liked to hit the sack as the dessert dishes were hitting the sink. "A tired faller is a dangerous faller," he'd say at least once a week, "an' when yer ugly as me, ya gotta get all the beauty sleep ya can!"

After some small talk about the cute new cook in camp, and the rising price of cigarettes, he popped the question. It seems his oldest boy

Photo: Larry Spencer

" . . . mountains everywhere and nary a road in sight."

needed a summer job in order to finish his final year at university.

"The kid's havin' a bitch of a time finding anything, so I thought I might break him in on the saw. He's into it, but I told him I'd have to run it by you first."

Photo: Larry Spencer

" . . . break him in on the saw."

What could I say? Billy MacBurney arrived in camp the next afternoon, and the following morning was sitting in the Jet Ranger behind his old man, waiting for me to blast them up the hill to their work site. Life happens fast out on the wild West Coast. One day, you're a college student crawling the halls of academia; a few days later, you're hanging onto a Stihl 056 in the back end of Knight Inlet, with mountains, mountains everywhere and nary a road in sight.

The young guy caught on real quick. From the computer tally sheets that track each individual faller and every cut he makes, it was obvious "Billy The Kid" had inherited his old man's talent for laying down the lumber.

Because every person on the claim has a two-way radio so we're all working on the same page, seldom is there any dead air. Give a guy a microphone, he becomes an instant disc jockey–as in, "GOOOD MORNING, IRON RIVERRRRRR!"

That can lead to some fairly amusing one-liners in the cook-shack at dinnertime, bordering on the ruder edges of family rivalry. Pilots firing shots at each other, like, "Hey, Cowboy–where ya goin' when ya get fired here?" Truck drivers asking the guy with the day's most flats if his brother-in-law still owns a tire store. Boom crew quacking off about the new "no fishing rods on the dozer-boat" rule. Mechanics bugging each other about the number of morning trips to the can, and the "hang time" on resultant vapours. Just your basic, general yahooism.

But the fallers never say much. As if what they do doesn't allow time for joking, or deserve it. In fact, Joe has a black and white decal on his old hard hat that says, "I HAVE SEEN THE EYES OF DEATH." Yesterday, I saw them, too.

I'd just finished my last cycle in the Sikorsky 61–"The Big Ugly," we call her–and was sitting in the office trailer doing my day's paperwork. Suddenly, the base radio started squawking: "MAYDAY! MAYDAY! Man down on C.P. 12, front quarter, bleeding bad!" It was Joe's voice, and he didn't sound too good. A minute later, I had the little Jet Ranger screaming up through the clouds, a handful of throttle and Joe in my headset, asking how much longer–pleading.

"Two minutes max, buddy!" I barked into my mike, suddenly realizing they'd switched quarters a few days earlier, and Billy was now working the front one. I banked down around the treeline to scope the scene. Not a pretty picture: Big Joe astride the butt cut of this ugly, old cedar slab, a rooster tail of sawdust belching out the back of his saw, and his boy pinned underneath, face down.

No more than five minutes had passed from the Mayday till I set down on the pad, and, by the

Photo: Cindi Collinge

" . . .'The Big Ugly,' we call her . . . "

time I hiked up to the accident, it was all over. I'd never seen a dead body before.

Joe sat there in the fresh sawdust, cradling the crushed head of his only son in his lap, and stared at me in the eerie mountain silence. His eyes looked like burned-out light bulbs. The late afternoon fog started rolling in, obliterating the camp down on the water. I knew in ten minutes we'd be stuck on the mountain for the night.

"Help me, Bobby," Joe whispered, softer than I'd ever heard him speak. I knew what he was asking. The main rule about a "fatal" is you don't move a thing until the Workers' Compensation Board guys give the OK, which means after their investigation. Joe knew it, and he also knew if we didn't go soon, we weren't going anywhere.

"I can't leave my boy up here overnight, partner," said Joe. I was in a very tight bind. If I went against the book and moved the body, I was leaving myself wide open for not only a suspension, but maybe losing my licence forever.

"I'll make it up to ya, Bob." His cracking voice sounded a thousand years old and a million miles away. By the time we wrestled Billy out over the felled and bucked and down to the machine and strapped him into the stretcher, the dreaded fog was closing in faster. We were lucky to get off.

Photo: Z. Olak

Big Joe in the back quarter.

Joe rode in silence, his eyes squeezed shut all the way down the mountain. Back in camp, the word was already out, and a crowd of guys was waiting at the pad to meet us. I set that baby down there like I was landing on my grandmother's grave.

I remember heading into my bunkhouse to write up a report, and looking back at the 206 rotor blades, still twirling above the engineers unfastening the stretcher. Joe was still sitting in the cockpit, hands clasped over his face, with his son's battered hard hat resting on one knee.

An hour later, a call came from head office, instructing us to terminate the shift three days early. Ten minutes after that, they called again, and the dispatcher came to get me from my room. This time, the guy who signs my cheques was on the line–most unusual. He told me the WCB was heavily pissed off, and that he was coming in on the Beaver at daybreak. I was forbidden to fly, as of immediately, but he was also bringing his lawyer from Vancouver, so I should zip my lip until they got here. Just before we lost our connection, he asked how Joe was doing.

"He'll be alright, Tom," I said, as the reception started to break up. I cranked the "squelch" button over a notch, but the crackle only got worse. I thought I heard a faint "Thanks, Bobby," but I couldn't be sure.

At midnight, there was a knock at my door. It was Joe, with a bottle of Scotch. As he handed it to me, there was no mistaking the connection. "Thank you, my friend," was all he said; then he was gone.

Later, I fell asleep thinking about a whole lot of things. Like flying in the fog, and my little boy, and the decal on Joe's hat–but mostly about the pilot who flew my old man, on his very last helicopter ride, down off the hill. ❑

Photo: Cindi Collinge

Photo: R. Cain

W.C. Smith is a phone-book moniker used by the semi-legendary Bill Smith, or "Smitty," from up Port Alberni way. At various stages of his life, he has been a logging truck driver, woods foreman, piano player, hell-raiser, poet, and finally, a doting grandfather.

These days, he and his wife, Elizabeth, are extremely active, though officially retired, and still a main stopover for any old loggers who want to chew the fat about the good ol' days.

A Chokerman's Dream

Adapted by W.C. Smith

I dreamed I'd died not long ago
And left behind all earthly woe,
But up in Heaven there was no diff–
I still set beads for dear old Cliff.
'Twas sure a wondrous place to be,
The camp was run by C.J.C.;
Hertel, he sat at Cliff's right hand,
Still "pushing" in that Promised Land.

Photo: B.C. Archives and Records Service

Our side-rods helped the King of Kings,
And gosh, did they look cute with wings!
They smiled so sweetly when they spoke
And treated logging as a joke.
We never had to strain our backs,
And drove to work in Cadillacs.
And sitting naked among the thistles,
Was Lana Turner punking whistles.

There were no sidehills in this camp,
And days were never cold nor damp;
There was no rain, no snow nor fog,
And sidewalks ran to every log.
There never was a thing to fix;
We all got paid each night at six.
Our commissary–grand and large,
And everything was free of charge.

Our chokers, you must realize,
Were made of gold, and strawline size.
And haul-back blocks, though strong enough,
Were weightless as a powder puff.
And twice in every working day
To help us pass the time away,
Dear Cliff in his old Chev would come
With twenty quarts of Navy Rum.

Photo: Z. Olak

I loved this place–I'd be there yet;
I left in sorrow and regret–
But I got fired; in my folly,
I used my halo for a Molly.
Cliff found out, and sad to tell,
Said, "Bill, you must go back to Hell."
Which sounded strange, perhaps because
When I woke up–that's where I was!

Adapted from an anonymous manuscript in the *Logger's World* (Dec. 1965) by W.C. Smith, Port Alberni, B.C. (1985).

Photo: Cumberland Museum

Damnedest Thing I Ever Seen

Back in the summer of 1975, I was tending hook on a Madill tower out in the Taylor River area, near Port Alberni. We had a pretty highball crew working for this one-side gyppo, and when the antique machinery wasn't being monkey-wrenched, the boys could produce like a bastard.

Photo: MB Franklin River

The Old Man was a good guy, as long as you didn't piss him off—which we obviously did on this particular morning. To this very day, all the fellas who were there wish we'd seen the ending to what was one helluva performance, or at least had a camera along. See, on the Friday afternoon before the July first long weekend, we'd moved the machine up to a new landing and parked it there, pipe still down in the saddle.

Come Tuesday morning, I was on my corner at five-thirty, the regular time, waiting for the crummy. No show. Ten minutes go by, fifteen, twenty, and now I'm starting to wonder if maybe we're shut down for heat, or maybe the bus came early, or maybe old Bob slept in, I don't know.

So I left my caulks, hard hat, and lunch pail by the telephone pole and headed into the house to phone the crummy driver. Five rings, six, seven—finally, a groggy voice answered, "Uuuh . . . hello."

When I asked what was going on, Bob yelled, "Fuckin' two-dollar alarm clock! Hang on, I'll be right there."

So, after a quick trip around town grabbing the crew, we finally got off the pavement and hit the gravel heading for the side, thirty minutes late. Bob was pouring the coals to the old Dodge panel, and by the time we fought our way up the last little spur, the tem-

Photo: Z. Olak

Photo: R. Cain

perature gauge was nudging two hundred. Hot.

We crested the last hill, pulled into the turn-around behind the boss's pickup, and couldn't believe our eyes. Somehow, the old boy had got the pipe standing up on the ram, fully extended, had both raising guys strung, and was just climbing back up to the winches so he could go ahead on 'er.

Jesus Christ! By now, everybody had their caulks on, so by the time the crummy got stopped, all the guys bailed out and made a beeline for the yarder. The Old Man didn't say a word. He reached down to give the engineer a hand up onto the running boards. Then he hiked to the turnout, jumped in his pickup, and sped off toward the dump.

For years after that, we often wondered: If the crummy had been another twenty minutes late, would ol' C.J. actually have tried to get that spar up all by his lonesome? Knowing the guy, I'd have bet money on it! ❑

The late Fred Lippon, a highball hooker around Alberni in the 1970s, always liked to sit next to a new kid on his first crummy ride. On arriving at the side, Fred would ask the greenhorn: "Mind if I have a peek in your lunch-bucket, Partner?" He'd study it awhile, and exclaim, "Hey! Alright! ham and cheese—my favourite! If you get killed this morning, can I have 'em?"

Gyppo in the Morning

Photo: B.C. Archives and Records Service

In his wonderful book entitled *Woods Words: A Comprehensive Dictionary of Loggers Terms*, first published in 1958, Dean Walter F. McCulloch from the School Of Forestry at Oregon State College defined gyppo as "a small logger; almost any woods job done on contract."

According to Colonel W.B. Greeley, the term began with western railroad building around the turn of the century. Several labourers from a construction gang would form a little co-operative with a few old tools and mules. The leader of this rag-tag team would then negotiate a separate contract for grading a station or two on their own. These ambitious entrepreneurs were said to be "gypping" the regular gangs by working all hours of the day and night and so established the name "gyppo" for a small contractor who survived by cutting corners any way he could. So many one-time gyppos have grown into medium-sized or large operations that the once-disparaging term has become much more respectable today.

Photo: Bob Dingsdale

Following is a fictional excerpt from a gyppo logging boss's diary:

DEC. 14, FRIDAY

Photo: Larry Spencer

4:30 a.m. Hard rain woke me up. Could mean too much snow up on The Switchbacks for the bus to make it. Guess I better take a run out there so I can call back and stop the crew, save that two hours' call time. They'd probably like a day in the hay, anyway, 'specially Friday.

4:50 a.m. Swing past the shop to pick up old haul-back blocks for traction. Should pump a drum of diesel into Tidy Tank in case the Cat is low on fuel. Damn bung is frozen tight. Skinned knuckles banging on screwdriver. Ouch!

5:10 a.m. Hit the gravel mainline. Trusty old pickup reminds me about getting new front shocks. Ten years, four motors, a couple of rear ends, too many frame welds—almost time to trade 'er in. Maybe next year!

Photo: Ted Maurice

5:25 a.m. It's coming down pretty good, and I'm not even at Ten Mile yet. If I headed back for town right now, I could tell the guys to stay in the rack, but if I can make it in to Turnbull Lake, I can hike up and get the Cat, if she'll start. Should've walked 'er to the bottom last night, but had to grab those friction blocks off that late bus—which reminds me, speaking of buses: I better encourage somebody else to get their "A" licence, seeing how Bob retires next year. Also must bring in landing saws for the Christmas shutdown—otherwise those damn four-wheelers will be out here prospecting.

Photo: MB Franklin River

5:40 a.m. Radio check. A-161 calling dispatch, does anybody read?" Damn it, I gotta get these tubes checked. Guess they'll see my pickup and figure out I've gone for the Cat. Hope there's a can of starting fluid up there—better grab a case tonight. Write it down.

5:55 a.m. Park pickup and start hiking. Wet snow not too bad if I get it plowed before it freezes. Flashlight needs batteries, I need a holiday, and the bills need paying. But the work is never done, or so they tell me.

6:05 a.m. Twisted my ankle on a bloody chunk of rock up on the second switchback. Better shave it off with a blast

Photo: Ted Maurice

of Amex first chance I get, before it gets a tire or two. Snow starting to thicken, flashlight completely dead—man, some days I sure love logging!

6:15 a.m. Finally got old D-6 to pop. Thank God for ether. Oh, shit! Forgot about this left steering clutch and how bad it pulls. Hope the winch isn't frozen or the bull-line fouled. Now I know what "zero visibility" means. What's that old saying? "Go ahead or go home!"

6:30 a.m. Looks like bus headlights down by the lake. If I dangle, I can meet them by seven o'clock at the bottom, throw on the hook, and Jerry can get them at least to the gate so they can hike in up the new spur. God, this is a rough-riding old bitch! Guess I should take a look at those rollers sometime . . .

Photo: Larry Spencer

6:50 a.m. Met crummy at my pickup; the bloody hooker's slept in again. That's twice in three weeks. One more, no more. If I go like hell, I can have him on the hill by nine o'clock, or I could spend the day on the rigging myself. Knew I should never have loaned my caulks to that green chokerman last week. He works two hours and then rides in with the first truck. Jesus!

7:25 a.m. Nobody home at Pinky's place. Heard his old lady's been chasing around and he jumped off the wagon again, so maybe I'll slide by the office and check the application pile. God, I hate hiring a hooker this time of year. Once they see we're flying four beads and what a goat show this is, they usually start phoning around, looking for a landing job.

7:50 a.m. Called a native fella who said Ray Seetcher is his uncle. That's a pretty good name to drop. Damn, was supposed to meet the Fisheries guy at the water tank at nine o'clock. He won't mind waiting; that's what those government guys do best. Wonder when we're due for another check-scale at the dryland sort?

8:15 a.m. Picked up overload permits to bring loader in tonight. Wonder about pilot car; guess I'll chance it, again. Got to renegotiate with the fallers on price for that back quarter. Should remind them to watch the ribbons a bit closer. Also need some boomsticks. Better ask Ted to work Saturday with me, change that old haul-back. Maybe get Butch to help me unload the machine on Sunday if the low-bed doesn't break down or get arrested. Wonder if we shouldn't drop the loader and walk across that old Taylor River bridge, just in case. Maybe just double up on the insurance instead.

Photo: MB Franklin River

Photo: MB Franklin River

8:30 a.m. Called the Better Half to suggest the weekend might be busy. She reminded me I promised to drive the kid's hockey team to Nanaimo, Sunday morning, in the big bus. Joe's boy plays, too; I'll ask him. Better swing by Port Machine Shop and see if those hydraulic hoses are ready for the drill. Wonder have I got time for coffee. Boy, the old town's sure busy this time of day. Guess all the bankers are heading to work! Geez, where did my morning go? ❑

Photo: MB Franklin River

Photo: R. Cain

Loggerisms #4

"The Cat doctor."
(A Caterpillar mechanic.)

"Cornbinder."
(Any International Harvester machines.)

"Dirty blue dick!"
(A substitute for real swearing.)

"He was drunker than a thousand dollars!"
(It once took three days to get this drunk.)

"It's time to dung out!"
(To clean out the trash, from cleaning out the stables in the old horse camps.)

"He's eating soup with a fork!"
(He's working, but not accomplishing much.) ❑

Photo: B.C. Archives and Records Service

My Worst Crummy Ride

"JEEEZUS CHRIST! WE'RE ALL GONNA DIE!" As a wake-up call, it needed work.

I've often felt that the only good thing about winter logging is the crummy ride is slower and longer: more time to sleep for the crew. And snow is quiet on the roof–lulling, not like noisy rain.

Uncle Bob, our bus driver, was a comfortable old piece of Company furniture. He'd been around for over thirty years, driving the crew for twenty, and not one major incident in all those miles. He only got flats on the way to work, or so it seemed, and the guys all loved him because he always got us there safely, even when he hit the same bloody potholes ten days in a row.

Photo: MB Franklin River

Old "Vice-Grips"–the nickname we called him behind his back because he always packed a pair with him everywhere–considered his crummy-driving job a position of honour, and he wore it like a war medal. That horrible morning, December 14th, 1981, seventeen loggers from Port Alberni were damn glad to know Uncle Bob was at the wheel, and if we had any chance at all, he would find it.

"For chrissake! Slow this fucker down!" In the middle of a morning snooze, the last thing anyone wants to hear is some moron beside him screaming. And nobody likes waking up to an out-of-control bus, fishtailing down Hydro Hill in a snowstorm.

So once everyone was awake–thanks to our panicky chokerman–we herded all the bodies to the very back of the bus, adding weight for traction. Wes Keeley, the senior truck driver, went right up and kneeled beside Uncle Bob, next to the gearshift. I guess he'd been in runaway situations before, because he didn't seem too excited, not like the rest of us.

By this time, we were rolling pretty good, the back end whipping around like a loose fire hose. Wes's slow, easy drawl cut through the darkness, as heavy breathing and the soft slap of windshield wipers made the scene hypnotic. "Ya still got 'er, Bob, but maybe we should keep the feet spinning, stay off them brakes–whaddya think?" He sounded like an air traffic controller easing down a crippled plane. Those two old-timers were zoned right in, totally focussed on the case and digging deep into a very large bag of tricks.

It was eerie as hell, looking out the side windows at pitch-black, while up front, the headlights cut a pair of golden arrows into the swirling snow. Half a dozen lighters snapped like a chorus of crickets as cigarettes scrambled to get lit.

"Y'know, Robert," Keeley said, calm as though sitting in his favourite bar, "if we can idle around that first little bend after the road sign, we might get lucky. Maybe ol' Numbnuts on the grader fell outta bed early this morning–ya never know with those road guys–and he just mighta scraped the runway clean already."

Wes had been driving logging trucks up and down these hills for fifteen years, and he knew the Highways Department always cleared the bottom half of Hydro Hill first. That way, if a guy lost control on the top half but still managed to get around the sign corner, he'd have the option of his brakes on bare pavement, eventually.

Photo: MB Franklin River

I thought I heard somebody whispering The Lord's Prayer; it could've been me. We were all wondering if that funky old crummy could walk around that hairpin at forty miles per hour without laying over, and it didn't take long to find out.

A few guys said later they felt the wheels lift off on Bob's side, but when we slid through that last curve still rubber side down and saw bare pavement in the headlights, man, that whole busload started howling like New Year's Eve.

Photo: Author's collection

After Bob got the bus stopped at the bottom and let everyone out for a leg-stretch and a round of high-fives, old Keeley put his arm around Uncle Bob's shoulder. "Partner, I been thinkin' . . . it might not be a bad idea to buy that grader guy a jug of rum for Christmas, just to keep him interested, know what I mean?"

Yesterday was the tenth anniversary of that hairy little bus ride, and last night the crew got together in Uncle Bob's rumpus room to hoist a few hot rums. We toasted the late, great Wes Keeley several times, and throughout the evening, each guy took a shot at imitating old Wes. Uncle Bob does him best. "Well, boys, it's like this. Some days you eat the bear, then one day . . . the bear eat you." ❑

A Logger's Sunday Morning

I awoke one day, as the poets say,
And my tongue had a new fur coat,
And a cavalry squad, I swear to God,
Was staging a drill in my throat.

Somebody with glue, though I can't say who,
Had pasted my eyelids tight.
And the whole damn bed was full of my head,
So I must have been out last night!

Original drawing by Bus Griffiths

Ole's Letter

For my sixteenth birthday I was given a well-used McCulloch hard hat, some monkey-face gloves, and a new pair of size nine Dayton leather caulk boots. Not exactly what I'd expected or hoped for.

But the real ass-kicker was the job setting chokers in the family logging operation. Again, a surprise, because my summer holiday plans were already well in place. This was my first head-on collision with "fallback position." Not to mention "worst-case scenario." Welcome to the real world, in that order.

But on Monday morning, July 13th, 1964, hiking up the sidehill with the rest of the rigging crew, I was given explicit directions toward a new purpose in life. The only words spoken to me on my debut day in the woods came from an old Swede hooker, Ole Nass, the boss logger. "Kit," he said, "do us all a big favour, OK? Yoost don't get kilt. Focks up perduction if I gotta pack ya down da hill, da ol' man get pissed off, maybe gonna can me, roon my whole bloody day!"

Photo: Ted Maurice

Sonny.

In twenty-five birthdays since, several of which I even remember, no single gift has changed my life so drastically as that first damn chokerman job.

Like most teenagers in the early sixties, my main concerns were accessing a car and getting a handle on my body's hormones–not necessarily

Don't get killed. Sounded like great advice at the time. Words to live by, so to speak, and they marked my introduction to a man famous up and down this coast as a true character. That summer, I was bitten deeply by the logging bug, and I guess working alongside Big Ole had a lot to do with it. Ole had a gift for getting wood to the

Photo: Z. Olak

landing.

"Ya yoost got to haff da right knack, boys," he used to tell us, "but ya got to haff dat left knack, too!"

The fellas in the crummy and the brass in the pickups all loved having Ole around. When Da Beeg Swede was on the scene, people were not only chuckling; they were also making money–lots of it. Seems the two go hand in hand–in the logging business, anyway.

Due to the ease of pulling the pin at one place and starting somewhere else a few days later, with no questions asked, guys like Ole bounced up and down the coast from gyppo to gyppo, eventually earning reputations as "camp inspectors." He once told me the thing he liked most about working in camps was being able to store up enough of a grubstake that some fine morning, he could finally yell in to the donkey-puncher, "Son, slack da main and call da plane–dis ol' huuk-tender is headin' downtown!"

Photo: Gord Barney

"Downtown"–that's what Ole used to call Vancouver, except when he was actually down there. Then, after a few drinks, he changed it to "Couverville," and finally, after quite a few more, he reduced it to just "The Couv."

Ole once told me, "Son, effry time I haff a goddamn drink, I feel like a new man, and yoost like a yentleman, I offer dat new man a drink!"

So eventually Ole moved on from my dad's camp, as I did, and we didn't cross paths again for several years. I heard on the guyline grapevine that he ended up in Toba Inlet, where he tried being married awhile. I looked forward to seeing him again so I could hear all the gory details firsthand.

Fast forward to December 1969. I was now a twenty-one-year-old truck driver, freshly snowed out of Nootka Island with a four-month paycheque burning a hole in the flap of my chain-drive wallet. Six of us from the crew, all single and whiskey-silly, had chartered a Goose and boogied down to The Big Smoke for the Creedence Clearwater Revival concert.

We were just snuggling in nicely at the old Ritz Hotel pub for an afternoon of some highball, beer-parlour logging, when I happened to glance across the bar, and there he sat, large as the hose on your fire pump.

"Son-a-my-bits!" Ole yelled. "How's da balls of yer feet, kid? Yeezus H. Kee-rist, ya need a haircut bad!"

So after he'd bear-hugged me half a dozen times and shook my hand dizzy, he ordered a plate of daffodils (a tray of beer) and suggested we "lick 'em up, RFN." Apparently, that had nothing to do with the Royal French Navy. Ole explained this was shortcut Swedish for Right Focking Now. After we'd assembled a large bouquet of dry daffies around our table, I decided the time was ripe to enquire about his marriage.

"Ole, you old stump-hooker," I finally asked, "now what's the lowdown on this rumour goin' around about you waltzin' to the wedding bells?"

Our whole table got deathly quiet. I figured I'd blown it badly, gone too far, too soon. Like maybe the sheets were still smoking, even though it was public knowledge she'd split with some young stud flunkey after only a few weeks with my old amigo. Ole drained his beer and fumbled in his coat pocket for another pack of White Owl cigars.

"Boys," he whispered, "a little advice. Always you should handle da booze; neffer let da booze handle you. Couple a drinks, sure, dat's OK. Couple dozen drinks . . . you could wake up married."

Photo: Ted Maurice

So we laughed about that, hoisted several toasts to some old bunkhouses we'd shared, swapped a few old lies with some new variations, and then it was time to go. I asked did he want to tag along to check out CCR, but he said he never paid to hear live music where he couldn't get up and sing with the band. Besides, he was expected somewhere else at eight o'clock sharp–or so he told us.

Just as our taxi full of party boys was pulling away for the Coliseum, I happened to glance back up the sidewalk, and there was Ole, struggling along with a cane in each hand, obviously in major pain.

"Pull over one sec!" I yelled from the back seat, and the cabbie jumped on the binders. By the time I got to Ole, he was white in the face, sweating like a pig in the cold night air. His giant chest was heaving like he'd just hiked to the back end packing a pair of haul-back blocks and three strawline extensions.

"What's goin' on here, partner?" I said, trying not to sound too panicky. I could feel myself sobering up real quick, and I didn't like it one damn bit.

"Relax, kid–we hired out to be tough, remember?" His voice sounded raspy, no longer disguised by the constant buzz of the rowdy beer joint. Looking away so I couldn't see his eyes, he handed me an empty matchbook cover. A phone number was scrawled inside.

"My landlady's kinda fussy on wisitors, so maybe giff me a call sometime–we go out for a beer an' a bullshat."

By now, the hack is honking triplet "Beep-BeepBeeps," my moron buddies are yelling at me to move my ass, and I know I've got to say goodbye. Big Ole Nass, who taught me how to "not get kilt" on my very first day in the woods, stood there in the freezing night, his old Stanfield pullover unable to button around his giant neck, just an old lion in winter.

He reached out a huge, gnarled paw to shake goodbye, and one of his canes clattered to the sidewalk. As I finally pulled my hand away and bent over to pick up his fallen stick, the last thing he said to me was, "Kit, yoost do me one small favour, OK? Don't forget what yer ol' man used to yell to the fellas effry mornin' in da landing. 'Boys, when she's too tough for effryone else, den she's yoost right for us.' You always remember dat, you'll be all right."

So anyway, the concert rocked big-time, and three days later I woke up in a motel in Chilliwack. They say it's an occupational hazard for loggers blowing into town after a stint in the boonies. The old compass needle gets spinning around, and a guy can lose all sense of direction, especially when he's guiding himself by the stars. I once knew a Newfie rigging-slinger who had a theory that the stars go to bed earlier in the city than out in camp, but, like I said, he was a rigging-slinger.

The following Tuesday, I decided to jump a bus into the city and check up on my old sidehill buddy. From the pay phone in the Alcazar Hotel lobby, I dialled the number on Ole's matchbook. A woman much too young to be a landlady answered: "Good evening, Vancouver Cancer Clinic."

Of course, this had to be a wrong number, so I checked–and, sure enough, the voice replied, "Yes, sir, that is the number you have reached. Now, may I help you?"

I felt like I'd been hit square in the headlights with a twenty-pound choker bell. "I'm looking for a Mr. Ole Nass," I heard myself whisper. My head began throbbing and the phone suddenly got slippery in my hand. Ten seconds of dead air. I thought maybe the switchboard had screwed up and cut me off. I prayed that's what happened.

"Are you a relative?" she asked finally, and in a much softer voice.

"Not really," I said, "but I know he'll definitely want to see me." More silence. What the hell is going on here? I thought I'd better grab a cab and hustle on down there, check out the program firsthand.

"Is your name George Spencer?" the girl asked gently, knowing it probably was.

"Yeah, it is," I said, by now quite shook up by these big-city surprises.

"Mr. Nass left a letter here for you, Mr. Spencer. Would you care to come and pick it up, or shall I mail it to you?"

So anyway, no one showed up to claim the body, and I eventually arranged to have his ashes sent home to his sister in Stockholm. Every year on my birthday–which was yesterday, incidentally–I still drag Ole's letter out, pour a stiff belt of Crown Royal, and read it over a couple of times. The last page is my favourite, where he says:

. . . and Women, kid, you just remember that's the best damn thing we got, like your mother or your sister. So always be a gentleman! You know, if the Good Lord invented something better, then He sure kept it for Himself, so now maybe He gonna let me have a peek . . . if I been a good boy!

And now about the Booze. Well, you just got to find that happy medium for yourself. Sometimes I found it too much, I guess, and now they say it kills me, but maybe I die ten years ago not even drinking, of something else . . . whose nose? My time here is up–that's OK by me. I had enough good times laughing for a dozen guys, and you should always laugh, kid, it's the second-best thing

So remember–Women are the best, laughing is second best, and laughing with Women . . . that's a Party. Always go to the Parties, kid, and stay as long as you can. You never know when it's gonna be the last one. Parties keep you young, but always catch the crummy in the morning!

And now about Logging. They call us loggers, but really we are just farmers in the woods–if we're gonna plant the crop again, for the grandkids coming along. All you young guys better get going on that real soon. I would love to hang around and help, but I guess I got to take this damn "promotion," whether I like it or not.

One last thing, young fella–you were always a good friend to an old Swede hook-tender, so thanks for that. You stay out of the bight. I'll have the blocks hung, and I'll be waiting at the back end. Don't hurry.

Ole Nass
A pretty darn good logger

P.S. It's a great life, kid, don't let it kill you. ❑

Photo: R. Cain

Loggerisms #5

"Just go up and eyeball it!"
(A preliminary guess by naked eye, not measured by instrument.)

"I need some new hand-socks!"
(The logger needs new gloves.)

"I'm going to yump the yob!"
(Jump the job. A Swedish logger leaving one camp for another, or for town.)

"How about a shot of YMCA whiskey?"
(A logger drinking lemonade or any other soft drink.)

Photo: B.C. Archives and Records Service

"He's yarded his last turn!"
(A logger who quit the woods for good, or died.)

"His brain is empty as last year's bird nest!"
(About as empty as it gets.) ❑

Photo: B.C. Archives and Records Service

The High-Rigger

Spectacular, daring, and now extinct, the high-rigger once had the most glamourous work in the woods. It was his job to climb the tree picked for a spar, lop off limbs on his way up with a one-man saw and topper's axe, and then, twenty-five or thirty feet from the top, loop himself to the tree with a rope. He would proceed to cut the top off, hanging on precariously while the top fell over him, the tree whipping and twisting in a twenty-foot arc.

A well-known high-rigger was "Flyin' Phil" Grabinski, who at one time held the northwest spar tree climbing record–scaling a hundred-fifty-foot tree in one minute and three seconds, lowering himself to the ground in just eighteen seconds.

"The first tree I topped was at North Bend, Washington, in 1922, and I was scared just enough to be very careful. I started up this tree that was about five feet through, so it took about twenty-two feet of rope to go around the tree and my body. The tree was one hundred eighty feet high and eighteen inches through on top, so it took about nine feet of rope up there. When I got ready to top, I took a good look all around. I wanted to know where I'd be after the top started over, and as soon as I saw all the spare rope I'd taken in as I went up the tree, I took it and threw it around the tree again. Having two ropes around me, I felt much better immediately. When the top went, it didn't even give me a thrill, and I thought this business was pretty easy—till I topped the next one!

Photo: Ron Dalziel

"Caesar" Scott - 1940

"It was a much larger tree, and as I was chopping, my body was swinging on the long, sharp spurs. They kept working steadily deeper into the tree, and pretty soon the damn top cracked, and instead of going away from me as I had figured, it came right at me. When I tried to get around the tree, I found my spurs dug in so deep I couldn't get them out. Right then and there I started thinking what a bad boy I'd been.

"It was a heavy top, and as it came over, it pulled the rest of the tree with it for ten or twelve feet, till the top was about forty-five degrees. Then it broke loose, and the tree swung back with me hanging on for dear life. The top fell straight down, so I learned it is impossible to get hit with the top of a tree up there.

"There are quite a few cases of high-riggers cutting their ropes, possibly one or two strands, at some time or other. A regular climber's rope consists of four strands of hemp over a soft wire core. Many people believe that the wire core in the rope makes it safe, but an ordinary blow with a tree-topping axe will cut it in two, as the wire has to be very soft to be usable in a rope.

"One reason why tree-toppers don't slice their ropes more than they do is because when we are topping a tree, we are chopping about three inches above the rope and striking lengthwise instead of crosswise. When we strike the rope, it generally rolls with the axe bit, and only one or two strands are cut. When that happens, we fix it before we go any further.

Photo: Z. Olak

"One of several things may cause a rope to be cut. One spur may pull out of the tree when you are in the middle of an axe stroke, or it may happen on a windy day. We always try to fall our trees with the wind. That makes us stand so the wind strikes us sideways and swings us around the tree. The tree is so small where it is being topped that our feet are right close together, and we can't brace ourselves as well as we should.

Photo: B.C. Archives and Records Service

"The real hazard of topping trees comes in windy weather, when there is danger of the tree splitting. If you get an undercut chopped into the tree, and a gust of wind comes along and blows the top the other way, it is liable to split the tree down the heart. If the tree-topper isn't fast about that time, and doesn't scamper down about twenty-five or thirty feet to where the split part breaks off, he is apt to be crushed against the tree by the rope and have his back broken. I once heard of a guy up at Quatsino Sound who had a big fir split on him, and when he grabbed on to the back half, the other side snapped back and pinched his thumb tight. He had to take his axe and buck off his thumb at the knuckle to get free. He was known up and down the coast as 'One-Thumb' Brown after that.

"The advantage of sawing over chopping in tree-topping is that in sawing, you just cut what is necessary to go through a tree with a kerf about an eighth of an inch wide. But in chopping, you have to cut so much dead wood. In a twelve-inch cut, your opening starts about twelve or fourteen inches wide, compared to a saw's one-eighth inch. With a saw, you are cutting steadily, while with an axe, only when the bit is in the wood are you doing any real work.

"The reason most toppers use an axe is because they have to have an axe along anyway for the undercut and the limbs. Then, too, the saw is dangerous to have hanging from your body when you chop a limb off, as it is liable to hit the saw as it falls and either jerk you up pretty hard or cause it to fly back and cut you–and it's much too heavy to carry through the woods, with belt, rope, spurs, and axe.

"The fastest job of raising a tree I ever witnessed or heard about was at St. Paul and Tacoma's Camp Number One. I was climbing, Cecil Coleman was the hooker, and Bill MacAbee was the foreman.

"We put up a one-hundred-seventy-two-foot spar tree in eight minutes, from the time it left the ground till we ran the mainline in, and the tree was tied with four guywires and perfectly straight. That might've been the hardest eight minutes anybody ever worked!" ❑

1990 LOGGING FATAL

WORKERS' COMPENSATION BOARD OF BRITISH COLUMBIA

CHOKERMAN

AGE: 19 – 1 YEAR EXPERIENCE AREA: VANCOUVER ISLAND

While logging in an area of steep slopes and unstable ground, a worker was caught and buried in a mud slide following heavy rainfalls.

Courtesy of Workers' Compensation Board of British Columbia

Photo: Z. Olak

Photo: R. Cain

Bar Vibrations #3

"I always have a bitch of a time sleeping the night before a move. If you ever seen a steel-spar come over, then you know what I'm talkin' about. She's not a pretty sight, layin' over on her side. Any hooker whose heart don't pound a little faster when that old pipe is balancing halfway up is either a pretty good bullshitter or a complete idiot! Bet I moved those tin towers a hundred times or more, and every goddamn time there was one moment when I thought, 'Jesus Christ, there she goes!' The whole idea of hooking one-inch cables onto stumps so a hundred-foot pipe can raise up in the middle of nowhere sounds pretty hare-brained, but I ain't lost one yet, touch wood. A lotta things have to happen in the exact right order. The key is finding real skookum stumps for the raising guys. If one of those guy-lines pulls out with the pipe up off the ram, your balls are in a very tight vise!

"The whistles are real important, too, 'cause if they ain't blown right, or if the engineer is too hung over to hear 'em properly, anything can happen. I don't know why, but most of the donkey-punchers I ever met were half rummy. I guess sitting in a steel cage all day with drums and lines clattering in your face, and chickenshit chasers wanting every second turn up-ended, would drive the average guy to drink. Over the years I met very few yarder operators who were strangers to the booze.

"I always liked to stand fairly close to the machine when she was going up or down, just to give the engineer some confidence that he wasn't alone. I wanted the guy to think I had 'er all under control, and all he had to do was run those winches whenever I gave the signal. What's that old saying–You can fool all the people all the time, if you know what you're doing! Maybe I'm just lucky, but you gotta be good to be lucky, partner." ❑

Photo: Z. Olak

Photo: Z. Olak

The Truck Driver

Perhaps the most sought-after job in the logging industry is that of truck driver, considered the glamour position. Every day in the cab includes danger, excitement, stretches of boredom, and lots of travel. Guiding a West Coast off-highway hauler loaded with a hundred tons of logs over some of the worst terrain through which roads have ever been built requires the nerve of a bullfighter and the skill of a jet pilot. One mistake and it's coffin-shopping time.

Although there are various heavy-duty equipment schools, air-brake courses, and personal tutors, nothing can totally prepare a budding truck driver for that first hands-on experience of a solo trip down the hill, fully loaded.

After securing the cinch lines, filling out the load slips, and checking the tires and slack adjusters, the driver climbs into the wheelhouse, sets his transmissions in the bottom gears–the basement–and hits the air horn three short blasts, informing the world that he is heading for the beach.

As he barks into his two-way radio, "A-200 leaving the tower," he informs all other vehicles on the Company network exactly where he can be expected on the road system. He is also expected to be aware of all oncoming traffic as they report their various positions, although the loaded truck always has the right-of-way.

Once underway, the truck driver must determine just how his load is riding–whether it is "on

Photo: B.C. Archives and Records Service

Photo: Ted Maurice

the pin" (perfectly balanced) or "side-heavy"—due to an under-achieving loader operator. The truck driver will apply this knowledge to the ever-changing lay of the road, which he has previously inspected on several dry runs with an experienced driver who has pointed out possible trouble spots.

Because the logging roads in the mountainous coast of B.C. are rough and steep as any in the world, a log hauler must have the brute power to battle up the hills and equal restraining brake strength or "stop" to keep the load from running away. Even today, some grades are so formidable that a Cat is needed at the top of the hill to hook onto the back of the trailer and winch or "snub" the load down to flatter ground. In extreme cases (usually at the end of a very long downhill grade, where the brake drums would be most likely to overheat and cease functioning) the roadbuilders will construct a "runaway lane." When a long, steep hill endangers his life three times daily, a driver's stress load is eased somewhat by the knowledge that the runaway lane is there, if needed.

With electric retarders, called "Jake (Jacob) brakes," which force the exhaust back through the motor to help slow the piston action, along with on-board computers that give instant information regarding every vital function of the truck's components, today's truck drivers are much advanced from the days when they were derisively known as "herders."

In their air-conditioned cabs, the logging truck drivers of the nineties are a direct link with that rich heritage of men who years ago sought and found a better way of moving the wood "from stump to dump." ❑

Photo: Larry Spencer

Photo: Bob Dingsdale

The gyppo arrives.

Photo: Author's collection

She owns it, she drives it.

Photo: Bob Dingsdale

Ted checks his lug-nuts.

Photo: Author's collection

Nancy checks her brakes.

Photo: Author's collection

The dead . . .

Photo: MB Franklin River

. . . and the quick.

Photo: R. Cain

Loggerisms #6

"They came down with a bad case of financial combustion!"
(Arson is suspected.)

"That guy is badly out of lead!"
(Suggesting that a person is not completely sane, describing someone who is not thinking clearly.)

"Give it a ten-gallon overhaul!"
(Paint it.)

"My bedspring relatives are coming to town!"
(My in-laws are arriving.)

"Pass me that log wrench, partner!"
(A logger asking for a peavey.) ❑

My Old Pal, Hal

For the last twenty years, I've made a regular visit to a man once revered around here as "the boss logger." He's old now–eighty-six, to be exact–and his name no longer rings the bells it once did, but this bothers him very little. "Boys," he used to say, "that fame bullshit is just a fancy paint job, and it ain't the cab that hauls the logs."

I was a week past my sixteenth birthday and green as old ham in a bachelor's fridge when I first encountered "the old man," as the crew called him–though never to his face, of course. At several inches over six feet tall, his woods nickname was "High Pockets," and I'll bet there are still hundreds of loggers scattered up and down this coast who easily recall the first time they ran into Big Hal Hertel.

Monday morning, the 27th of June, 1964, my shiny new Dayton caulks boldly announced that a rookie chokerman was making his grand debut. While the equipment operators climbed into their cabs to warm up their machines, the rest of us stood around in a huddle, sipping coffee, making small talk. Down below our landing, a cloud of dust scurried up the switchbacks.

I knew by the way the crew got immediately quiet that the battered pickup slamming to a halt just twenty feet away carried the foreman. No one else in his right mind would need to be anywhere in such an ass-shaking hurry, unless they were running the show. The man certainly knew how to make an entrance, and the jaunty tilt of his hard hat left little doubt he was "the brass."

I found out later that Big Hal took secret pride in going through a pickup every two years. In fact, the local Ford dealer had an understanding with the Company not to waste time

Photo: MB Franklin River

Photo: Z. Olak

bringing Hal's trucks around for a trade-in. "Deetroit sure don't build 'em like they used to," he'd crow, while a welder or mechanic laboured over his latest disabled vehicle. We often wondered if that was a veiled reference to his tank-driving days in World War Two.

As the diesel engines of both yarder and log loader roared to life, exhaust stacks belching twin arrows of black smoke before settling into a slow, gentle throb, Hal Hertel, boss of bosses in the country's largest logging company, approached me with outstretched palm. "Son," he said, squeezing my hand in his giant meat-hook, "if yer gonna make any mistakes, make 'em now, 'cause I seldom fire guys on their first day. But tomorrow you got no excuses, know what I mean?"

The way he said it—obviously his standard spiel for new guys—sounded like a cop reading a prisoner his bill of rights. As he released his industrial-strength grip, he winked down at me. "Just remember, young fella—if we don't get that round stuff, then there ain't gonna be no flat stuff." Then he walked away, chuckling.

Big Hal Hertel was a man truly ahead of his time, and the lessons he taught us young bucks back then were far more valuable than money. "It's war out here, boys, and people get killed in war—so keep your head up an' your ass down, 'cause funerals piss me off!"

He had safety on the brain long before that became fashionable, and saved his worst freakouts for anytime he saw people working dangerously. Years before, he'd buried his younger brother, and it haunted him. "Jeezus-fucking-Christ, how many times do I have to tell you pimply-assed cackzickers—ya don't drop a guyline to yard past

Photo: Gord Barney

Photo: Larry Spencer

diamond lead!" But as mad as he'd get, he would forget just as quickly, and everyone knew it was because he cared so much. He cared about other things, as well. "I don't know why those bastards in head office won't listen to me," he'd complain. "I tell 'em we're logging too damn fast, an' we should be replanting, but they all say that's not my problem. Well, fuck them, 'cause it IS gonna be my problem when my grandkids can't get a job and are living in my goddamn basement!" And this was way back in the sixties.

I suppose my eagerness to learn any job on the claim landed me in his good book right from Day One, and whenever I asked about something new, he'd say, "OK, but I'm only gonna show ya twice. Once so's ya learn it, and once so's ya don't forget!"

The day he retired, he drove around to all the sides in his area so he could shake hands with every guy who worked for him. I was hooking on the new grapple-yarder on a steep little show at Sproat Lake, and here comes the old bird, huffing and puffing right out to the back end. I was busy fighting a hang-up, using the whistle, and he got right up behind me and listened awhile to me yelling and cursing the situation. Then he put his hand on my shoulder. "Son," he said, "it's only a goddamn log—don't let it ruin your day."

Photo: Gord Barney

Today, some of the things that old gaffer said still crawl across my mind periodically. Like every time I use a piece of two-by-four, or thumb through a newspaper, or drive past a new building going up, or whenever I hit the corner store for a couple of rolls of bum-wrap.

A schoolteacher friend thinks it would be a great question for her grade five history class. "Kids, for twenty-five points and a chance at the bonus question, tell me what came first—the round stuff or the flat stuff?"

Maybe she should ask an old logger, while there's still a few around. Like my old pal, Hal. ❑

The Runaway

This was supposed to be my thesis on "The Zen of Truck Driving," and I wish certain people had lived long enough to see it in print. It is their legacy; I just wrote it down for them. My intent was to offer an appreciation of the truck-driving life, and to document the lives of some artists I met along the way–including the late Al Jaworski, who was lucky to live through this story. For those who don't believe that God has a twisted sense of humour, Al survived one of history's great truck crashes, only to die of cancer a few years later. He told me this story on his second day out of the hospital, and his first day back in the truck. So keep the soft side down, partner, and I'll meet ya at the water tanks, over.

Photo: B.C. Archives and Records Service

Photo: Z. Olak

First of all, you don't have time to be scared. When the shit hits the fan, you don't even think about the panic button. Everything happens so goddamn fast, your brain shifts direct into "automatic"–kind of like natural instinct, I guess.

See, inside the cab, everything happens at a certain pace that you control, usually. But when she finally gets away, it's like someone hit the "fast forward" button on your video machine. Just like that; everything's a blur immediately. You can lay awake lots of nights thinking about it, wondering how you'll react if and when, but until you actually ride your first runaway, nobody, I mean nobody, can tell you how you'll handle it.

Every load of logs is different, every truck has its own personality, and every driver has a different panic level. So any time you climb up in that cab and aim toward the beach, you better have your act together. If you don't have at least a small shot of fear mixed into your game plan, one of those big cedar peakers could end up in the coffin factory, with your name carved right on it!

You talk to any trucker who lives through a runaway, and he can tell you for half an hour every tiny detail of a ride that probably took two minutes, max. Two of the longest minutes he'll ever experience!

It was first thing Monday morning after the Labour Day long weekend. I was in this old Mack junker backed under a brand new Skagit 555 on rubber, with Charlie Allen at the controls, one of the very best log loaders on the coast.

Photo: MB Franklin River

I remember thinking how gentle old Chas was with the grapple, setting the bunk load on there with a touch so soft I didn't spill a drop of coffee. With most guys, you didn't even pour till they were halfway up the stakes. Then I remember Charlie calling me on the two-way radio after he'd blown "shave and a haircut," the signal that loading was finished.

"Yeah, Al, maybe ya wanna throw the binders on right where ya are. I didn't have much lay for that top one." It took several tosses before I got the cinches over the peaker, and, by the way the trailer tires were bulging, I knew I had a real jag on.

I also remember having to drop both sticks right in the basement to climb the little adverse up out of the landing, which should have been my first clue. Usually, I walked out of there in second and under, no problem. Twenty minutes later, I had that little Cummins 220 in the big wheel wound up tight along a stretch we call "The Freeway," just before the start of the last, steep switchbacks.

Photo: Gord Barney

See, back in the sixties, everybody had a "go ahead or go home" kind of attitude, and the philosophy toward shutting down a wood buggy for preventative maintenance was, "If it ain't broke, don't fix it!" For some reason, brakes were never given top priority, and the line most truckers used to the loaders was, "Don't worry about it. You load 'er; I'll haul 'er."

I was just starting down the first switchback, and my air gauge was falling way too fast for the ten pounds of trailer brake I'd applied. I flipped the lever all the way off, and the needle headed straight for "Zero." The load was so damn heavy, it began pushing the truck, and I had to keep gearing up so too much compression didn't blow the pistons completely through the engine block.

I could smell burning brake lining as the scenery out my windshield was coming at me lots faster than usual. Water poured on scalding brake drums gives off the same odour as burning old tires and camel-dung together. A unique aroma.

Now warning lights were flashing all over my dash, my tach was pinned to the red line, and it hit me–Jesus Christ, I've got a runaway! But what I remembered most were the words of the Old Man, the first time I ever rode with him: "Son, if she ever gets away on ya, don't jump too soon. Sometimes you're better off stickin' with 'er, ridin' the damn thing out." So that's what I did.

I woke up to the sound of water splashing over me and the hissing of steam. The glare from the welder's cutting torch hurt my eyes as he worked away on the mangled metal pinning me inside the cab. Two mechanics with backpack cans played a steady stream of water down on us to keep the burning to a minimum. See this blue scar on my neck? That's my souvenir.

But what I remember most is hearing my brother's voice, after it started to get dark.

"Man, I tell ya, some guys have all the luck. Look at that–he didn't even break his goddamn thermos, for chrissake!" ❑

Photo: Gord Barney

The Back-End Telephone

Back in the ungodly hot summer of 1964, all logging operations in the Alberni Valley had been shut down tight for the entire month of August. By late July, the hazard sticks were so goddamn dry, we couldn't yard or load after nine o'clock in the morning. Even the old-timers agreed it was a bit warm.

The Forest Service had slapped a total closure on the woods, and all the local beer parlours were jammed with thirsty loggers waiting for the call to go firefighting.

By about the middle of August, the arrow on the highway hazard board pointed to "extreme." Everyone in town was getting antsy, figuring it was a matter of time before the whole bloody Island exploded.

Finally, on Labour Day weekend, there was a sprinkle of rain over the coast, out Ucluelet-Tofino way. That brought enough overnight recovery to allow my old pal Jack McQuillan to crank up a few sides down at Millstream Timber. But Jack had one small problem.

I guess most of his rigging guys had scattered when the dry spell hit, and he needed a few days to round 'em all up. Sunday afternoon I got a phone call, and it was Jack.

"Cliff," he said, "we've got the OK to head 'er back tomorrow morning, but my goddamn hooker went out for a trip deckhanding on his brother's gillnetter, and I heard they took along a case of whiskey–you know the story–so I'm in a bit of a bind. Seeing how you guys are still shut down, I was wondering if I could borrow Ron Dalziel for a few days, just till we get things up

Photo: Z. Olak

and running."

I told him that would be no problem. Not only was Ron a hotshot hook-tender, but he probably wouldn't mind getting out of this pressure cooker for awhile.

"Where exactly are you logging at, Jack?" I asked, thinking if it was on this side of Kennedy Lake, maybe Ron would just commute from Alberni.

"Cliff," he told me, "we're so goddamn close to town, the crew could walk to work. You know where they stuck that new phone booth, right there at the Ukee-Tofino junction? Well, if you were making a call and the door was open, you could piss on our tail-blocks!" That sounded pretty close to me.

About an hour later, Ron phoned. "Cliff," he said, "old McQuillan wants me to go and yank a few pecker-poles for him out there near Tuff City. Says he's already booked a Steely Posturepedic for me in the Ukulele Lodge, so if I'm not high-centred in the hoosegow, that's where you can find me for the next few nights."

The next morning, I was sitting in my office, listening to the nine o'clock news, waiting for a five-day weather forecast, when the phone rang. Phone calls in fire season always make me jumpy. I grabbed it on the first ring.

"Collect call for anyone, long distance from Ucluelet," the operator said. "Will you accept the charges from a Mr. Ron Dalziel?"

Oh, Jesus! Immediately, I started thinking the worst, like maybe he missed the crummy, or had a union hassle, or . . .

"Yes, I'll accept the charges, operator."

"Hello, Cliff," Ron said, sounding quite far away. "How's every little thing in there this mornin'?"

Photo: Ted Maurice

"Everything's OK in here, Ron. What's the problem out there?" I figured he'd probably launch into some long, drawn-out tale about the usual Monday-morning wobble.

"Oh, there's no problem here at all, Cliff," he said. "I just wanted to let you know how good these guys have got it out here. Sure nothin' half-assed about this outfit. They even got a goddamn telephone at the back end so the boys on the beads can call home for lunch, if you can believe

Photo: Author's collection

From time to time
Ron still calls in—collect, if he can.

that! Well, Cliff, it sounds like the rigging's comin' back, so if there's any messages for me, my number out here is 726-1020. Talk to ya later!"

He must've held the receiver out toward the approaching chokers, because I could hear the damn bells banging and clanging together in the background, like a herd of cows heading for the barn.

I often wonder what he would've done if I hadn't accepted that call. Knowing Dalziel, he probably would've borrowed the coins off some poor, rubbernecking tourist watching the action from the roadside. He still loves to remind me about this outfit he worked for once that was so damned classy, they even had pay phones at the back end. To this day, I chuckle every time I drive by that phone booth. ❑

This story was told during an after-hours gathering of contractors and owners in the Madill hospitality suite at a recent Truck Logger Association (TLA) convention.

There was a very bad accident on the Squamish Highway involving a Jamaican, a Yankee, and a logger. Shortly thereafter, they all arrived at the Pearly Gates.

"What's goin' on, boys?" asked St. Peter.

"We all just got killed in a car wreck," they answered.

"Well, you look kind of young to be dead," St. Peter replied, "so here's what I'm gonna do. I'll grant you all one wish—whatever you want."

After a brief discussion, they all wished to be back down on Earth. St. Peter thought a bit and then replied, "OK, boys, for twenty-five dollars cash, you can all spend a little more time back on Earth."

The Jamaican took out his wallet, gave St. Peter two tens and a five, and BANG, was back in his old neighbourhood immediately. His buddies quickly gathered around him and asked, "Willie, what happened, mon? We thought you was dead!"

Willie smiled and replied, "That ol' St. Pete is one cool dude, boys. For twenty-five bucks, he let me back here awhile!"

His friends looked puzzled. "What happened to them other two guys with you?"

Willie chuckled. "Last I heard, the Yankee had ol' St. Pete down to seventeen dollars, and the logger was lookin' for a co-signer!"

Logging in the Thirties

An Evening with Pete and Ina

Pete Thorpe-Doubble is now seventy-seven, and still sharp as a Marlin spike. With Ina, his beloved wife of over fifty years, he operates a fifty-acre farm near Duncan, B.C. When visiting them, be prepared to help put up some hay, go riding his herd of horses, stay for dinner, or perhaps assist in finding the bottom of a jar of Mister Walker's blackest.

Pete ran a trucking outfit around Victoria back in the days when, as he puts it, "Men were men, women were women, and toilets were half a mile outside."

Photo: B.C. Archives and Records Service

I first met Jack Cronin in the late thirties when I was hauling out near Kemp Lake. He was over eighty then, living in an old shiplap cabin with knotholes you could see through. He had been a teamster back in horse logging days, before machinery marred the woods. Anyone that ever mentioned his name always added, "the best damn teamster on the Pacific coast," so I assume he was.

Skidding logs with horses and oxen was big business back in Jack's day. The skid roads had short logs set in the ground crossways, several feet apart, for the logs to slide on. The lead end of the logs were sniped off so they wouldn't hang up on the cross skids, and the skids were continually greased with whale oil from a "pig."

Photo: Alberni Valley Museum

The pig was actually a hollowed-out log, much like a small canoe, which was dragged along behind the turn of logs with a tub of skid grease in it. The tongs, dogs, and chains were loaded into it, to be dragged back to the woods.

Loggers have always used indecorous terms for certain operations, sayings you would never use in front of your maiden aunt. The man who greased the skids was known as a "pig fornicator," but not quite in such delicate language.

The hiring halls in the cities had a blackboard out on the sidewalk listing all the available jobs in the camps. Apparently, some dear little old ladies read the board from time to time and registered a firm complaint with the city fathers. From then on, whenever a skid greaser was needed, the board read:

Wanted: one P. F. man.

❑

Photo: Pete Thorpe-Doubble

In the Dirty Thirties, I hauled logs for Yat Fine, a happy old Chinese gentleman. Most of his crew were Chinese, with the exception of a few key men and the whistle-punk. Steam "donkey-punchers" were never Oriental, even in Japanese or Chinese logging companies, because the government would not issue a class four engineer's ticket, or "donkey special," to Orientals.

One of the westerners that hooked and rigged there was a nervous, panicky kind of guy, stuttering and stammering all the time. The chaser in the landing was a Chinese named Lum. One day, the rigger was up the spar tree and wanted something sent up on the pass line. Lum tied it on so tight that the rigger couldn't undo it when he got it up the tree. He yelled down in absolute frustration, "F-F-For Christ's sake, Lum, c-c-can't you see I've only got t-t-two pair of hands?" ❑

Photo: Cumberland Museum

One day during the summer, I went to Yat's house to get a badly needed drag. There was a terrible noise coming from the next room, an awful commotion of wheezing and hacking. I thought someone was being throttled. "What in hell is that?" I asked him.

He just laughed. "That my poor wife," he said. "He got fay heber." ❑

Photo: B.C. Archives and Records Service

Another time, Yat had sent four Chinese fellas down to the booming ground to stow logs. When I arrived with my load, some wild shouting and ky-yi-ing drew my attention to the brow log. One guy had fallen in and was thrashing around, yelling for help, while the other three just stood there, screaming, but didn't lift a finger to help. I got out there along some boomsticks as fast as I could and managed to get a pike pole out to the drowning man before he went under for the last time. After I got him ashore, I asked the three onlookers, "What in hell is the matter with you bastards?"

"No can do! No can do!" they all shouted at me, yet they all seemed very grateful for what I'd done.

A few days later, when I asked Yat why nobody had helped save the guy, Yat replied, "No can do."

It turned out that the Chinese people believed if a person saves another from certain death, it becomes the responsibility of the rescuer to look after the victim for the rest of his life. Apparently, none of them wanted that millstone hanging around their neck. ❑

Photo: Pete Thorpe-Doubble

Talking about Oriental boom men, when I hauled for a Japanese outfit up at Mud Bay, our dump often went dry at low tide. Then it would plug up before enough water came to float the logs clear. Clearing the dump was often done at night, or whenever there was a high tide. One morning, I went down with a load, and the dump was still plugged from the day before. I asked the Japanese boom man why he hadn't cleared the dump yet.

He stared at me a few seconds. "How I can do it?" he said. "All dry water stop." ❑

Photo: B.C. Archives and Records Service

When I look back on it all now, I don't feel too good about how the Japanese and Chinese were treated in the early days of logging. I worked for some of them, played with lots of them, and just like any other group, there were some good guys and some jerks. But we all had lots of laughs, and we sure got a helluva pile of wood! ❑

Loggerisms #7

"He was packing an equalizer!"
(A club, marlin spike, or other fighting tool.)

"That guy can really flip the chips!"
(A logger who is fast and expert with an axe.)

Photo: B.C. Archives and Records Service

"He folded up his face!"
(A flap-jawed logger who finally quit talking.)

"What a godawful pile of goop!"
(Any dirty, sticky, smelly combination of grease or oil.)

"He's got a half-nelson on the payroll!"
(A man with relatives among the Company bosses.) ❑

1990 LOGGING FATAL

WORKERS' COMPENSATION BOARD OF BRITISH COLUMBIA

LOGGING HANDYMAN

AGE: 46 12 YEARS EXPERIENCE AREA: MID-COAST

After unloading equipment on a floating wharf the truck travelled up the wharf ramp. The left-hand ramp stringer broke at mid-span. The truck plunged into 10 feet of water, drowning the operator.

Courtesy of Workers' Compensation Board of British Columbia

Photo: Z. Olak

How to Catch a Crummy

Catching a fish is hard work, catching a cold is miserable, and catching a crummy is a little of both. There comes a sad morning when every logger has to catch his very first crew bus. When that crummy door closes behind him, the rookie learns quickly that an entirely different set of rules governs this strange, new world he's entered.

Rule Number One is simple. Seating is arranged by the weird dictates of age, job status, and sheer physical size–the back seats are reserved for new guys. Besides, it's much warmer up front, and there's more leg room.

Rule Number Two is harsh. "Siddown and shaddup, till we decide if yer worth talkin' to!" In fact, it may be several days, sometimes weeks, before the green recruit is even introduced around. No sense learning a guy's name if he ain't gonna make 'er. But in the meantime, the new logger can catch up on Company folklore from the older guys up front.

"Say, Bob, remember that time Harry had the runaway in the little Hayes, when he met ol' Numbnuts coming up the hill, and they both dove over the bank? Was that up Branch Twelve, or were we out the back end of Canoe Creek by then?"

"Well, for one thing, it was Jack at the wheel, and he was in the old Mack before she burnt up the second time, and he was comin' down that horrible little spur off Branch Five-O-Five, right where we laid that culvert that got the new Fisheries guy all hot and bothered, so it must've been 1962 out in Ash River, if I recall correctly."

"Oh, sure, now I remember!"

But while the veterans discuss the price of a new haul-back, lack of performance by the grade

Photo: Author's collection

Photo: Larry Spencer

Uncle Bob

crew, or the hassle of filling out time cards, the young guys in the back have far more pressing concerns.

"Hey, Rick, you know that waitress in the Tally Ho, the one with the great ass? What's her name again?"

"Forget it, dude–she's shacked up with some cop. By the way, you still got that spare ticket for the Stones concert on the ninth?"

"Say, home-boy, you know where a guy could score some decent herb for the weekend?"

"Partner, you keep smokin' that shit, it'll make yer dick small. Look what happened to mine!"

Sometimes the assortment of individuals thrown together on a crummy resembles a mobile Johnny Carson show. If there were some way to capture the giddy glee percolating inside a crummy bound for town on a payday Friday, a person could bottle it, flog it, and retire to the Bahamas, or maybe even Winter Harbour. And, as on Carson, there are certain topics of conversation that should be avoided: politics, religion, and the sexual inclinations of the hook-tender's wife. It's a very small space, and a long ride home. And for some reason–Murphy's Law, perhaps–most flat tires and breakdowns occur when the crummy is heading home, rather than in the morning, when a delay would be more appreciated.

One of my favourite crummy stories is told by Andy Bohun about his first experience with West Coast logging. It seems he'd arrived in Nanaimo, fresh out from Ontario, with high hopes of becoming a B.C. logger. Eventually, he got a phone call from the employment office, and was given a number where they might be hiring. The ensuing conversation went something like this.

"Hello, I heard you might be looking for a logger?"

"Ya got a good hard hat and some sharp caulks?"

"Yessir, I sure do!" Lie Number One.

"Ever hooked behind a skidder?"

"Sure have." Lie Number Two.

"You're hired. Catch the crummy tomorrow mornin', ten to six, outside the Queens Hotel."

Now Andy was nervous. He immediately phoned his uncle, who had logged in various camps around the coast before a back injury landed him a WCB pension. Andy's uncle patiently explained about hard hats, and caulk boots, and how to hook the chokers onto the bull-hook of the skidder.

Andy was about to hang up when he finally remembered one last question. "Oh, yeah, Uncle Ted–one more thing. What will I need to catch a crummy?" Dead silence for a few seconds.

"Well, son, for a case of beer they usually slow down long enough so's you can jump on."

Andy says he felt pretty stupid the next morning when the driver told him, "Sorry, son, ya can't bring a case of beer on the crummy. Company rules, y'know."

Like old "Spike" Marlon, the donkey-puncher from Sooke, used to say, "Boys, there's a lotta ways to catch a cold, an' a lotta ways to catch the clap, but there's only one way to catch a crummy. Ya just gotta be there early!" ❑

Photo: MB Franklin River

Late one night, a robber came up to an old gyppo and said, "Give me your money, or I'll blow your brains out!"

The old logger just smiled. "Go right ahead on 'er, kid," he said. "I can always log without brains, but I can't log without money!"

Photo: Z. Olak

The Wreck of the Old #9

Another W.C. Smith Poem

Photo: Alex Spencer.

Before Smitty's Time.

Besides poetry, Smitty drove truck for twenty-five years before promotion to logging foreman at Sproat Lake Division. He finally departed after speaking his mind to the Supervisor–in detail and most explicitly. He drove the pickup into camp and rode shanks pony to town. He went to work for Coulson-Prescott Logging right after this, and that's when I got to know him.

He isn't writing about someone else's experience; he, himself, was the lucky guy who got to go along with old #9 on her wild ride.

It was Monday morning, and old "9" and me
Went up to see Miles at the skidder tree.
"You're first up the hill," said Loukes with a shout,
So I boarded old "9" and broadened her out.

Photo: Ted Maurice

Photo: Author's collection

Al West and Smitty

And she pulled like hell, for well she knew
We had to be first up Branch 22.
She went up those hills just like the devil
Well above the two-thousand-foot level.

Till dead ahead we could plainly see
A silhouetted form—the skidder tree.
Miles was waiting with a comical grin;
He seemed to sense that trip of sin.

I put ol' "9" underneath the boom,
Moved over in the cab to give Hank some room.
Hank and I talked about this and that,
While Pard was loadin' her wide and fat.

At last he came down and said he was through—
"You've got thirty hemlock, and you're wooded, too!"
I climbed from the cab feeling proud and fine—
"Hell, that's just a jag for old '#9'!"

Again Pard gave us a big, wide grin,
And I hollered, "Get back there, and pull out my pin!"
It was about seven-thirty when we left that side
For what was to become a helluva ride.

Photo: W.C. Smith

Actual Photo: Old #9 at the end of the poem

Along the flat, then we checked to see
That the water in "9" was running free.
So we broke over the hill all feeling fine:
The load, and me, and old "#9."

The load was heavy; we found that out
As I felt for the brake pedal and flattened it out.
There was lots of air, but she wouldn't slow
And then old "9" began to go.

I grabbed the emergency and pulled it up tight,
Dynamited the trailer, but she was in full flight.
So I said to "9," "Baby, I sure can see,
Where old Dad's going to leave your company."

Smitty:"Two weeks! Can't you have her ready by Friday?"

Photo: W.C. Smith

I patted her dashboard as she started to wail,
Then through the air I started to sail.
I landed safely as God had willed,
And I watched old "9" damned near get killed.

I looked over the bank and nearly cried,
For I was sure old "9" had instantly died,
But give us two months and ten thousand bucks,
And again she'll be one of MacMillan's best trucks.

And I know she's waiting to be fixed like new
So we can go back to the show on Branch 22. ❑

Smitty mugs up on the fly.

Photo: Bob Dingsdale

Smitty's biggest load: 124 tons of very heavy hemlock.

Photo: W.C. Smith

Photo: R. Cain

"Goin' down the road with everything wide open but the tool-box."

W.C.Smith

1990 LOGGING FATAL

WORKERS' COMPENSATION BOARD OF BRITISH COLUMBIA

LOG-LOADER OPERATOR

AGE: 51 – 20+ YEARS EXPERIENCE AREA: VANCOUVER ISLAND

While operating the log loader on a self-loading logging truck, the main rotation supporting pin broke in two. The operator was thrown to the ground and suffered fatal injuries.

Courtesy of Workers' Compensation Board of British Columbia

Photo: Z. Olak

Photo: Ted Maurice

Bar Vibrations #4

"Thirty-nine years is a long time working for the same goddamn outfit. I been here longer than the new foreman's been alive, for chrissake, and next month they're giving me the gold watch treatment, waving me out to pasture! Y'know, I'm starting to feel like the old Peggy Lee song, 'Is That All There Is?' Sure, it sounds pretty good, dragging in seventy grand a year with eight weeks paid holidays, but what happens now? I guess that's all she wrote, Louie.

"My biggest mistake was getting too damn good at my job, and they froze me there. Once the brass saw I could load logs like a bastard, that's where I bloody well stayed. Right from when I started, second-loading on an old SJ-4 tong machine, I could see some guys were real artists, the way they'd dance down a log and toss them big steel tweezers with perfect timing so's they spread open and fell around the butt, almost like ballet, y'know? I figured I could learn that, too.

"I remember the very first day I grabbed 'er by the face: Monday morning after the Labour Day long weekend, September 4th, 1953. A few rangitangs on the crew didn't make it back to camp after a bender in town, so the Old Man asked me did I want to take a shot at the controls.

"Man, I was up in that cab right fucking now, before he changed his mind. Two hours later I'd stacked my first load of logs on a wood-buggy, and there were four more lined up, waiting for a jag. Three years after that I broke the Company record by throwing on eighteen loads in one shift. I tell ya, Jack, I thought my shit was roses!

"But that's the Catch-22 about the logging business. You have a talent for a certain machine,

Photo: Z. Olak

soon you're running it better than anyone else, producing like a bastard, and then the brass start counting on those numbers, so they dangle a few perks and bonuses so's you don't look around, and there you stay. For thirty-nine goddamn years!

"Every year you climb another notch on the seniority list, and after too many funerals and retirement parties, eventually you've got the longest whiskers on the claim.

"I know hindsight is 20-20, but, God, looking back on some of the stunts we pulled, we were nuts. Monkey-wrenching till midnight on second-hand junk that was worn out by the time we got it, laying in mud trying to jump a track back on, greasing every weekend, forever splicing rotten old mainlines, freezing your balls off in winter, walking that old Dominion dipper-stick down goat trails with nothing to stop me but the rocks I ran into–Jesus, those were the days!

"Now they tell me I gotta pack 'er in. What am I gonna do–sleep till noon, go fishing, belly-ache to the old lady, listen to my kid rave about his new machine? Can you imagine–an on-board maintenance computer, air conditioning, and a stereo tape deck for lunch time relaxing. He says their new general manager even gives pep talks to the crew about personal 'stress-management.' It sounds like I'm getting out just in time.

"One thing I do know, if he's gonna break his old man's load record, this outfit's gonna have to get smaller trucks or bigger logs. Funny thing about this old logging game though, I miss it already, and I'm not even gone yet!" ❑

Photo: Ted Maurice

Best Excuse I Ever Heard

Back in the late sixties, we had quite an interesting mixture of characters on the crew, mainly because we were considered a gyppo. MacMillan Bloedel seemed to have some rather selective hiring policies, so guys came to us as a last resort, before heading out to a camp.

See, working for a small outfit back in those days was considered the bottom rung on the logging ladder. But, small as we were, we gave everyone a fair chance, kind of like the Oakland Raiders football team. It didn't matter what a guy looked like or smelled like or dressed like; if he could do the job, that was all we cared about.

As a result, we ended up with ex-cons, longhairs and rummies on the wagon (and a few off the wagon). We had white guys, native guys, Germans, East Indians, Japanese, you name it–all of them rejected at other places for various reasons. But they all got along great, and they sure knew how to get wood, which is always the bottom line for a logger.

Of course, with a bunch like that, we often had to make allowances for unforeseen emergencies–usually on a Monday or a Friday, for some reason. Every Sunday night, I expected at least a couple of calls telling me about a doctor's appointment or a court appearance or the fairly regular "grandmother's funeral"–usually good for once a year, with some guys.

Sure enough, one night our family was sitting around the tube, watching the Ed Sullivan show, when the phone rang. Immediately, the kids started taking bets on whether it was going to be a doctor, dentist, or lawyer's appointment. It was Charlie Mickey, our rigging-slinger, who hadn't missed a day's work in over three years.

Photo: Gord Barney

"Sorry to bother you, Pat," he said, "but I don't think I can make 'er tomorrow."

"Geez, Charlie," I asked, "what's the problem?" I thought I heard sirens in the background–not like the party noises I often heard in these situations.

"Well, Pat, my goldarn house is on fire, and the truck just got here, so I guess I better go. I'll probably see ya Tuesday."

When I got back to the living room, the kids all wondered what I was chuckling about. I told them it was just a business call.

The following Wednesday, in the weekly newspaper, there it was, dead centre on the front page. A picture of two firemen up on a ladder, putting out Charlie's chimney fire.

Over the years, I've heard a lot of wild explanations for missing a day's work, but a guy who asks for time off so he can save his house from burning down has got to win any company's "dedicated employee" award. Yep, that was far and away the best excuse I ever heard. ❑

Photo: Z. Olak

Helicopter Logging

Men have been harvesting trees many thousands of years, but only logging with helicopters since 1972.

Helicopter logging is the ultimate, most expensive way of extracting timber, bar none. But in many situations, it is the only way. In areas too ecologically sensitive: along streams or estuaries where building roads and skidding logs would be destructive, or in hanging valleys economically out of reach for conventional logging systems, heli-logging is the only alternative.

Past logging practices, particularly during the railway era, left large volumes of isolated, high-value timber that cannot be reached without damaging the newer, second-growth forests. All of these areas are potential heli-log sites.

There are a few key requirements to ensure heli-logging is a paying proposition: assessing timber values is most important; the "turn-around" time must be kept short; the payload high; support costs, too, must be kept to a minimum—without sacrifice of safety.

Wayne Coulson, Bob Hawthorne and Gary Collinge have developed a computerized, aerial-harvesting system that is unique in the logging world.

Up until 1987, the old system of heli-logging had chokermen hooking conventional wire-rope or nylon chokers into a hook on the end of a two-hundred-foot cable attached to the helicopter. The turn was flown to a drop site, lowered, and the chokers released electronically. The chokers were then gathered in slings and flown back to the logging site by a smaller helicopter. In this method, the accumulated stress on chokermen and pilots was extremely high.

The solution was a hydraulic grapple invented by Coulson Forest Products in Port Alberni. Today they are still the only logging company in B.C. to own and operate heavy-lift helicopters used exclusively for timber harvesting and fire suppression.

Chief pilot Bob Hawthorne says, "The big helicopter companies often use their heavy-lifters contracting for different logging companies. They'll pull the wood and then leave to do something else, so the work is sporadic, hit and miss. Work a month, then no logging for a couple of months. With Coulsons, it's a whole different attitude as the 61's are just a tool to log with—a flying yarder. We run as an aviation division of the logging company, and we obey all the rules of Transport Canada and the WCB, but we've always got a steady supply of work. I'd been at Okanagan (now Canadian) for ten years when Cliff and Wayne approached me. When I looked at their machine, it was like it just came out of the showroom. Compared to what I'd been flying, it was a real beauty. Then I went into the boardroom and Wayne explained his vision of the grapple. My dad was a high-rigger and a logger all his life, starting on steam donkeys, so logging is in my blood."

Wayne Coulson says, "If a company has a Tree Farm Licence but doesn't utilize the twenty percent of timber in their tenure they can't get with conventional logging, then that wood should be made available to more creative entrepreneurs who have figured out how to get that difficult wood. There is no way the industry can continue taking out only the valley bottoms. In B.C. we've got thousands of logged out drainages that still contain good

Photo: Cindi Collinge

merchantable timber. Utilizing one-hundred percent of the wood supply has got to be the goal of every company, big or small. If one day all we have left is helicopter wood, the government will not get paid as much stumpage as conventional logging pays, and the province will be strapped for cash flow."

Helicopter logging requires a very different approach than conventional logging. Compared to grapple yarders or steel spars, which rely on their enormous power to move the logs, a helicopter is a delicate machine. A "chopper" does not like heavy, dead lifts, so various techniques have been devised to "launch" the logs.

Hawthorne says, "Coming from the aviation world, with all it's idiosyncrasies, into the logging world was quite a learning curve, I tell you. We were creating something new, and we both had our own ideas how to go about it. Working hours was one case where we disagreed. They couldn't understand why, after six or eight hours in the cockpit, you were tired. Fatigue? What's this fatigue stuff?

"People don't realize that when you climb inside a 61, you don't have a second of relaxation. You're going up and down, lots of noise, vibrating all the time, total concentration–and it gets to you. Come for a ride and see what you think. So there was an education process both ways, and after a few rides they now understand all the variables, especially weather, wind, fog, sunlight and rain on the windshield. Loggers had to learn why yesterday we got eighty meters an hour while today we only got seventy. We had to explain how today there were adverse winds so we couldn't make our approaches the same way. Every setting I look at, I have to figure out what's the prevailing wind so I'll know where to put the landing. Because we work closely with Fisheries and the Forest Service, their input allows us to consider every

Photo: Bob Hawthorne

option for minimal impact to the environment, and then we choose the flight pattern that'll be the quickest and safest.

"I know that on the West Coast inlets, the winds flow in during the day and out at night, and from about October through the winter the prevailing winds are southeasters. So once I know if the wind is coming around the hill or over the hill, or back down the inlet, I can then decide where the landing should be.

"A blow over 20 miles an hour makes things really dicey. The ship starts to roll over and you have to point your nose into the wind, you can't come in with the wind on your tail. The machine just doesn't like it. It quits flying.

"The 61's don't like heavy rain, either. Big rain will disintegrate the tail rotor blades like you've taken a ball-peen hammer to them. If you keep flying in heavy rain, the leading edge of your main blades begin to look like Swiss cheese. On the tail rotor blade, the problem is with a strip of metal along the trailing edge that holds the thing together. The rain gets under that metal and eats it, pries it wide open. Hail is definitely out. If it hails, you're parked.

Photo: Bob Hawthorne

"A 61 is like an old '63 Chev pickup. Its a queen to fly, a real sweetheart. When Sikorsky designed this thing in 1958 his systems were so far ahead of their time that the others are just catching up now. It's built tough, and you feel good flying it. You get in, hit the starter, and away you go.

"Training on a 61 to be a logging captain takes at least a thousand hours before you get up to production speed. You learn that on the job. You can start pulling wood off the hill after a hundred hours but to get all the little techniques and to learn how to read logs and know the best way to launch them off the hill takes time.

"With heli-logging, the condition of the wood when it arrives at the sawmill is much better than with conventional logging. Instead of dragging it through the slash and mud, shattering against stumps and ripping the bark off, it comes out clean, so the value of the wood you get with the

Photo: Cindi Collinge

helicopter offsets some of the extra costs.

"When you come into a landing, you set the logs down. The 61 doesn't like a sudden release at all. The Sikorsky itself weighs thirteen-thousand three-hundred pounds and when you set down eight-thousand pounds of logs you want to release that load gradually. If you pickle that thing from the air, its like POW! You can bang you head up against the top of the cockpit. Sometimes you misjudge, or the wind screws you up and the choker goes slack, then tightens up again–that's called a hit–it'll tear you right out of your seat. It does airframe damage, too.

Photo: Cindi Collinge

"When Sikorsky originally designed these things, they were meant for troop carriers. Over the years we've made frame modifications to beef up the sling points in the belly; more beams on the inside. If he saw what we do today with his helicopters, he'd turn over in his grave. But after so many years logging, we know where the stress points are. We know that the tail boom is going to get cracks in it and where they're going to happen, so every four hours the engineers are checking these points. We also know that transmission mounts crack all the time, so we're always

Photo: Cindi Collinge

watching. Nothing surprising ever happens to a 61. She'll let you know what's going to go wrong.

"It's a very expensive piece of logging equipment–two million dollars just for the machine, let alone all the support equipment. You don't just haywire it together if something goes wrong!

"The on-site crew, not including managers and office staff, is two pilots and two engineers per

Photo: Cindi Collinge

Photo: Author's collection

day, and we try and work seven days a week. If we extend the flight hours to over 8.8 hours, we go to three pilots. The engineers have a long day. They start an hour early to preflight the machine and average about three hours of maintenance after the flying is done for the day. The pilots arrive half an hour before 'wheels off' to start it up. No one complains, because that's the way the industry is.

"Every time we land for fuel, which is every seventy-five minutes, the fellow who's done all the flying will get out of the left seat and switch over to the co-pilot's seat. The co-pilot gets into the captains seat. You couldn't fly that machine and look out the left window much longer. In the right seat, you're pushing the throttles, monitoring the weights and doing the book-keeping.

"When I'm having a good cycle, I can feel the adrenalin start to flow. You get into a rhythm and production really starts to happen. You hook up, turn, launch off the hill, land it, and you get higher, and the better you go, the more you get into it. It reminds me of a feeding frenzy of sharks when everybody gets into the rhythm of a great cycle. Then, as soon as you have a couple of aborts it breaks the rhythm, and you've got to start over again."

As Cliff Coulson says, "It's been a real learning experience for both sides, as the logging industry and the aviation industry speak totally different languages. But we've worked at it and have reached a real good understanding. We've got two very successful teams working together here, and I put that down to the people involved and their abilities to communicate with each other."

Wayne Coulson explains, "We are a logging company, not a helicopter company, so we've really pushed at blending the two together. To sit a loader operator with twenty-five years experience down beside a pilot who has worked all over the world was a learning experience for both. Moving some of our older employees–fallers, loaders, truck drivers–around the country with the Aircrane crew has opened everybody's eyes. We've done jobs from Alaska to California, and we're proud of our system. We operate in minutes. How many turns per hour, how many minutes per turn–so we've got tight control and total accountability to offer our customers. With our technology in radios and computers that govern our fall-

Photo: Cindi Collinge

ing system, we gain ten to fifteen percent total value on the wood, starting right at the stump. A lot of our decisions are based on the knowledge of our employees, and there is no substitute for practical people who are open to new ideas. It takes lots of long nights brainstorming in the board-room, but it works itself out."

For the past four years Rita Toews has been the Operations Manager for this company that has fought fire and evacuated villages in Manitoba, logged a huge pine-beetle killed area in California, fought fire in Idaho, and logged and rescued several downed aircraft in British Columbia. One machine is currently under contract to the Klawock Heenya tribal group in Alaska, which keeps Rita's telephone busy seven days a week. A mother of two teenaged boys herself, Rita considers the three big Sikorskys her other "babies."

"We are all really attached to the helicopters," she says. "When something goes wrong, we all feel it, from the pilots to Paul Ryan and his engineers to all the support staff in the office. It's a full-time challenge to keep all the machines healthy and flying, and we put our team up against anybody's. We have a worldwide parts inventory on computer along with an Airborne Data system. At the end of each day, the maintenance crew enters the number of hours flown. The computer will update every component on the ship, and then print out the part numbers of whatever needs immediate attention. The computer also warns in advance of future needs–say an engine overhaul in one hundred hours–and advises when to order those parts so they arrive close to the time they're needed. We've worked closely with the MOT on this system and they think our program is tops!"

While the open sky of helicopter logging has been there to explore for the past twenty-five years, once again an independent "gyppo" logger, Coulson Aircrane Ltd., picked up the challenge, and flew with it. In 1991, they received world patents on their hydraulic, flying grapple. ❑

Epilogue: A Chokerman in Paradise

Toba Inlet was a great place for dreamers. Electricity was homegrown, men wore their demons like medals, law was barely a rumour, and the major topic of conversation was WOMEN. Welcome to logging camp, Sonny.

"It's God's country, and He can bloody well keep it!" said the Trans-Mountain Air pilot, in a bad Richard Burton accent. How come these guys spend thirty years over here and still sound like they just tumbled off the Queen Mary, I wondered? Moments later, he cranked the prop on a museum-reject Beaver and aimed us straight for The Hole In The Wall, just out of Campbell River. Ten minutes later, I was praying out loud.

"God, Partner, Buddy, let's make a deal. You get me back on solid ground without swimming, we'll sign a brand-new contract!" Fortunately, He did–but, stupidly on my part, we didn't.

Nineteen years old, my first week of camp tucked safely under my suspenders–I figured I was New Jack Shit. Big-Time Stuff. Shortly thereafter, a light flashed on in my throbbing little brain, and a strange voice whispered, "Maybe you should write some of it down, before you forget."

But back in the sixties, real life got in the way of good intentions, and no one sang it better than the Animals:

When Ah think of all the good time
that Ah way-sted . . . havin' good times . . .

Photo: R. Cain

Photo: Z. Olak

That summer, the gang of hooligans from Bunkhouse Two at L. & B. Timber Company set about carving another name into the history of hairy-assed logging legends. My bed was just inside the door. "Youth is wasted on the young" never had a better example than those rangitangs I fell in with.

Merely surviving another day on the mountain called for serious celebration and "cocktails till ya drop." Big-time boogie in the bunkhouse, Louie—take no prisoners, check your brain at the door. Work like dogs, party like men.

In those days, "the future" meant later on—tonight. A case of Hiram Walker's finest was considered good stress management, a planeload of party girls chartered in from Downtown was "importing talent," and rules were for fools and wise men. We were looking for another catagory.

But the delusion of thinking you are living a movie is a Hollywood snow job. When you answer the cold call of the woods, there's no guarantee of a happy ending. Wake up and smell your Stanfield, pal.

Back in the days of Elvis and the Beatles, the logging merry-go-round spun so fast and crazy, no one ever considered the worst-case scenario. Falling off, getting left behind—not a chance! Well, guess what, sports fans? And we never saw it coming.

My friend Mike, a psychologist who makes big bucks telling rich people how messed up they are, explained it this way. "First, the east got settled—Columbus and gang. Then, as the hordes arrived and things got crowded, the more adventurous types pushed on. Eventually, word filtered back that the Prairies were happening, and the West got wild. Men became cowboys, women found related employment, and before you could say Wyatt Earp, the party was jumping. Soon, these wagon-train types ran flat out of prairie and smack into trees, so they parked their horses and became loggers. In fact, if John Wayne was still alive, he'd be doing movies with hard hats and power saws instead of Stetsons and Colt 45's."

Photo: Ken Soderlund

Photo: Larry Spencer

Photo: Author's collection

Sometimes I agree with Mike only because he's a doctor, though he does make damn fine wine.

Ah, if only this book had turned out the way I first envisioned. *A Chokerman In Paradise*, I was going to call it–just a kid full of dreams, lying on a rumpled cot in a shack jammed between the rocks and cedars at the dark end of Blowhole Bay. Unfortunately, this book missed a few things.

It missed the roaring silence from the crew as an eager, peach-fuzzed punk climbed aboard the crummy his very first time, bound for places known only by private code. "Ya can start out in Ash River, up Branch Five Hundred on that first little spur past the slide, hookin' behind the Cat."

And it forgot the mumbled goodbyes from another crew when that same logger finally pulled off his boots and stood in his driveway, watching the crummy pull away, heading for the neon. Ah, payday Friday.

Yeah, this book would've had all that and everything in between, if I'd done it right. It was a chunk of life that Rimbaud called *A Season In Hell*, and I doubt he even set a choker. Sitting here tonight, melancholy as a homesick Irishman, I realize those days are gone forever. But what really hurts is knowing those years we took for granted will soon be stuffed away in the archives of memory, abandoned like attics of buried treasure. Forgotten, unless someone writes it down.

If this book had nailed everything it should've, you'd hear the metallic crunch of new caulks biting into a frozen buckskin on a crisp October morning, when the Old Man was paying the boys a nickel a log after two hundred, and our nickname was the Tin Hat Track Team.

Somewhere in these pages, too, you might stand and gaze awhile at the million-dollar view from the top of Catface Mountain, where the air was so nose-pinching pure it was sweet pain just to breathe, and the hooker used to tell us, "When yer a B.C. logger, boys, every day's a picnic!"

Photo: Z. Olak

Photo: Author's collection

It should also have Johnny Oberg, who went from whistle-punk to high-rigger to watchman in his sixty years of logging (see picture above), telling a story, with every line so lip-biting funny in his thick Swedish accent that guys would start laughing in all the wrong places, and Old John would have to tell the story twice, just so everybody heard it once. "Son-uff-my-bits, boys, dat wuss one tuff blutty show, ah can tell you dat for shure, Gottdammitt to hell, anyways!"

And if this book got it all, you'd hear the sound of a haul-back pulling apart, a thousand feet of pissed-off steel python whipping through the tail-blocks, slicing anything in its path like a razor through warm butter. Then the eerie, awesome quiet on the hill before the rigging boys crawled from their hiding spots and took a head count. "Man, we were bloody lucky today!" Yeah, maybe—but you got to be good to be lucky, partner.

I wish this book had caught those lunchtime trading sessions when an old-timer would ask,

Photo: Larry Spencer

"What do I get for a chicken sangy?" and some poor rookie would offer up a salmon sandwich or lettuce and tomato, only to discover he'd just made a deal for basic baloney on white crusts. Welcome to the outfit, Slick—enjoy!

If this book had all I dreamed it would, there'd be that Monday morning at the marshalling yard when the owner told the alky foreman to park the pickup and hand over the keys, right in front of the whole crew. "I should break your goddamn nose!" the booze-hound yelled. They stared at each other for about a minute, then the Old Man calmly slid both hands into his pockets and said, "Well, ya better make the first one good, son, 'cause I get the last one."

And what about the first time that Indian rigging-slinger said to me, "Looks like we're gonna have to squaw-hitch that big cedar, partner." I'd heard that ugly word before, but never from a native person, and I remember how shocked I was. A few months later, I asked him about it over a bottle of rye. Ray just smiled. "Partner," he said, "When I was a young boy, the men from our village had a saying: 'If you saw my squaw with my eyes, you'd know why I want my squaw.'"

I could ramble on forever about everything that got left out, but, as Ole Nass would say whenever I complained about a tight log or snarly bead, "Quit yer bitchin', kid—it's good enough for who it's for!"

There never was a flight plan, really, and the mistakes I've made here were more fun than trouble. A foreman I once worked for up at Todd Inlet put it this way: "Son, you bullshit your friends, and I'll bullshit my friends, but let's not bullshit each other, OK?"

To accept that the stories were better than the storyteller was a bullet I had to bite. Where I come from, we dance with who we brung. This is sort of how it happened; I wish you could've been there. It was quite a ride. ❑

Barry Coulson, 1992
Port Alberni, B.C.

Photo: Ken Soderlund

Photo: R. Cain

Glossary

A-frame: Giant A-frame structure on a large log float, used to harvest timber along the steep sides of lakes and inlets. There are few A-frames left today.

Arch: Steel arch on tracks towed behind a Cat in order to raise the front end of logs up off the ground for easier yarding.

Backcut: Cut that the faller puts in a tree opposite the undercut.

Back end: The outside perimeter of a setting; the far end of a logging claim.

Bag or **Bag boom:** Boom sticks coupled together to encircle and enclose loose logs, popular in helicopter logging.

Barber chair: When a tree being felled splits, leaving a slab attached to the stump resembling a chairback.

Barge: A type of scow for transporting logs or machinery.

Bead: Slang referring to a choker.

Bell: Steel clasp that slides up and down a choker, into which the end knob is placed to "choke" the log.

In the bight: A bight is a bay, or sag, or bend in a wire-rope yarding line, particularly as it lies on the ground or flops around. When straightened out by a sudden pull, the bight will lash any man or anything inside the bow. This is a very common cause of logging accidents, frequently fatal. “Keep out of the bight!” is the First Commandment in the woods.

Booming: The act of stowing and sorting logs in the water.

Boomstick: A sixty-six-foot log bored with a four-inch auger or hole at either end. This enables several sticks to be joined together with boom-

Photo: Larry Spencer

chains, forming a bag enclosure.

Brow log: A large log used horizontally at log dumps to protect piling and to keep trucks in place while dumping.

Bucking: The act of sawing felled trees into measured lengths.

Buckskin: Slang for a log with no bark.

Bull-bucker: Supervisor of the falling and bucking crews.

Bull-cook: Usually a broken-down old logger who does odd jobs around a camp.

Bull-hook: Large hook on the end of the winch cable on the back of a Cat.

Bullpen: Enclosure where logs are dumped and held in the water for sorting.

Bung: The plug or stopper in a barrel or drum of liquid.

Bunk: The steel beds of a truck and trailer upon which the logs are placed for hauling. Hinged stakes are fastened to either end of the bunk to allow logs to be stacked.

Bunk-bound: When logs were loaded too close to the back of the cab, then the truck and trailer would bind up on sharp curves. Compensator bars and sliding reaches fixed this problem.

Butt-rigging: Hooks and swivelled connections joining the mainline, haul-back line, and chokers.

Cat: Short for Caterpillar, the company that makes the track machines (Cats) so popular in the logging industry.

Cat-skinner: A tractor operator, usually a roadbuilder.

Caulks: (pronounced "croks") Steel nails once driven into the soles of logging boots. Today, they are threaded and replaceable.

Chaser: The worker that unhooks the chokers in the landing at the bottom of the spar tree.

Cherry picker: A mobile loading machine used to pick up logs along a right-of-way where timber is close to roadside.

Choker: A length of cable or wire rope, usually seven-eighths of an inch or one inch thick and about twenty-two feet long, with knobs at either end and a sliding bell in the middle. One end is attached to the butt-rigging, while the chokerman pulls the choker to his log, loops the other

Photo: Ken Soderlund

end around the log, and fastens it into the bell.

Claim: Timbered area staked out for future logging.

Cold deck: A pile of logs yarded in to a spar tree and left for future loading or possibly continued yarding.

Conky: Unsound timber that is soft or rotting inside.

Crown fire: When a fire climbs into the tops of trees and travels with the wind. The most dangerous type of fire.

Crummy: A bus or vehicle that transports the logging crews to work.

Cull: Logs with no merchantable value. Disparaging slang to describe a useless logger.

Cunit: Measurement term for one hundred cubic feet of wood.

Deadhead: Partly submerged log.

Dipper-stick: Slang to describe old roadbuiding shovels that were often converted to log loaders with tongs.

Donkey: Slang for the machine with the drums and cables that sits at the bottom of the spar tree and yards the logs into the landing.

Donkey-puncher: Slang for the man who operates the yarding machine.

Dozer-boat: A powerful little boom-boat used to push and stow logs in a bullpen.

Eye-splice: An eye or loop spliced into the end of a cable or wire rope.

Fairlead: A series of blocks or rollers on a machine to guide the lines so they spool properly onto the drums. Used on the heel boom of shovel loaders and also on the top of steel spars or yarding towers.

Flume: A wooden chute filled with water used to float logs down to the salt chuck, mostly in the early days.

Flunkey: Someone who helps the cook around the cookhouse.

Fore-and-aft: A road of logs laid end to end, used in the early days.

Gilcrest jack: Used by hand-loggers to roll or push logs from the hillside down into the salt chuck. Very cumbersome; its nickname was "Kill-Christ" jack.

Goat show: Extremely steep logging situation.

Grapple: Large, steel set of mechanical arms that clamp around a log to yard or load it.

Grapple-yarder: Huge machine for yarding logs that does not require use of chokermen.

Ground lead: Old style of logging before spar trees, when logs were dragged along the ground.

Guyline: A strong cable stretched between a

Photo: Pete Thorpe-Doubble

solid stump and the top of a spar tree to ensure the spar remains upright.

Hang-up: A situation when a log being yarded becomes fouled in a root, in brush, or behind a stump.

Haul-back: The line attached to the butt-rigging that pulls the mainline and chokers back to the next turn in the woods.

Haywire: Describes anything in logging that is poorly run or inefficient. A term of derision.

Herders: Slang to describe truck drivers that don't do any maintenance or monkey-wrenching. Derisive.

Highball: Describes a logging operation where speed is king.

High-lead: Method of yarding toward a landing using a spar tree to lift one end of the logs into the air so they don't hang up so easily.

Jackpot: A jumble of criss-crossed logs. Often occurs when a load is dumped in shallow water.

Jag: Slang for a good-sized load of logs.

Jagger: A steel strand from a worn choker or cable. They hurt!

Landing: An area to which logs are yarded for loading onto trucks.

Low-bed: The truck and trailer used to transport equipment.

Mainline: The line used to yard logs. It has a larger diameter than the haul-back line.

Marlin spike: A sharp-ended, tapered steel tool used for splicing the strands of cable or wire rope.

Molly or Molly Hogan: A round link made from a single cable strand used to connect cables or to secure shackle or block pins in place.

Monkey-wrench: The act of fixing broken equipment.

Patch logging: A method of leaving strips of timber for seed source or firebreak and clearcutting the areas in between.

Peaker: The top log on any load, usually a large one.

Pike pole: Long, aluminum tubing with a sharp, spiked end, used by boom men to push logs around in the water.

Pipe: Slang to describe a portable steel spar.

Pilot car: The vehicle preceding large equipment being moved on a two-way road that warns oncoming traffic.

Peavey: A wooden tool with a steel spike on the end and a hinged cant-hook used to roll logs.

Photo: B.C. Archives and Records Service

Peeler: A high-grade log suitable for plywood or veneer.

Powder monkey: A roadbuilder with a blasting certificate, trained in the use of dynamite. Sometimes called a "Boart King."

Pre-load: To make ready a load of logs on a spare trailer with hinged legs so the truck does not waste valuable time.

Push: Slang for the woods boss or superintendent.

Raising guys: The two guylines that must pull the spar tree up off the ram and into an upright position.

Ram: The hydraulic tubing on a steel spar that raises the pipe up far enough out of the saddle so the raising guylines can take over.

Reach: The steel tubing part of the trailer that connects to the truck and contains the air and water hoses for brakes.

Rigging: The assortment of swivels and hooks that fasten between the mainline and haulback line into which the chokers are attached.

Rig up: To set up a spar tree for logging.

Road: The yarding path of logs.

Roll: A method of placing the choker around a log so it will spin in the desired direction to

Photo: Alberni Valley Museum

avert a hang-up.

Saddle: U-shaped cradle on a steel spar that holds the pipe in place when the machine is travelling.

Scab: Term of derision for someone who works through a strike or in a picketed area.

Schoolmarm: A tree or log that forks into two tops.

Second growth: The crop of trees that are seeded or regenerate naturally after an area of old-growth is logged.

Second loader: A tong-man that works with the head-loader.

Setting: An area where felled timber can be yarded and loaded with machines. A place where logging occurs.

Side-rod: Someone who bosses over one complete logging side.

Slash: Slang for any logged-over area.

Snag: A standing, dead tree, which is a dangerous fire hazard.

Snare: Slang for choker.

Snorkel: An extension to a machine boom, used for both loading and short yarding.

Snoose: Copenhagen snuff, the logger's substitute for tobacco.

Spar tree: The tree topped and limbed for use in high-lead logging in conjunction with a donkey.

Squaw-hitch: Slang to describe how a choker is looped around the largest possible portion of the end of a log.

Stag: To cut off logger pants between the knee and ankle.

Strawline: A light wire cable with an eye and hook on each end, used to pull the haul-back out through the tail-blocks and to change roads. One length, two hundred to three hundred feet long, is called an "extension."

Stumpage: The fee paid by loggers to government as a tax for the right to cut standing timber.

Tag: When two or more chokers are joined together in order to reach a stray log that is far off the road line.

Tail-block: A haul-back block strapped to a stump so the cable can run through it and pull the butt-rigging and mainline back for the next turn.

Talkie-Tooter: The name of the company that invented a small transistorized radio device that can blow loud whistle signals at the yarder or allow voice contact between the hill crew and the yarding engineer.

Tight-line: To put tension on both haul-back and mainline so the butt-rigging and logs are raised up out of the brush.

Tommy Moore block: A small block used to

Photo: B.C. Archives and Records Service

run strawline.

Tongs: Large, hinged tweezers with sharp, steel points, used on early log-loading machines.

Turn: The logs yarded in one trip of the rigging.

Undercut: A large, horizontal V, cut in the front side of a tree to control the intended direction of fall.

Whistle-punk: In the old days, a person who gave the signals from the rigging crew to the yarding engineer.

Widow-maker: A loose chunk or limb hung up in the top of a tree that can come down with no warning.

Wire axe: The device to cut cable when splicing must be done. At one time, just an old axe stuck into a log with the cable laid over it and the sledge hammer applied with gusto. ❑

Photo: B.C. Archives and Records Service

This book is dedicated to:

Bus and Margaret Griffiths,

Pete and Ina Thorpe-Doubble

and my family.

" . . . the tourists come and go,
but the loggers seem to stay.

Alma Edwards, 98
All-Sooke days, 1992